Methods in
Cellular
Immunology

Rafael Fernandez-Botran, Ph.D.
Assistant Professor
Department of Pathology
Division of Experimental Immunology and Immunopathology
School of Medicine
University of Louisville
Louisville, Kentucky

Václav Větvička, Ph.D.
Assistant Professor
Department of Pathology
Division of Experimental Immunology and Immunopathology
School of Medicine
University of Louisville
Louisville, Kentucky

CRC

CRC Press
Boca Raton   New York   London   Tokyo

616.07
F363m

Library of Congress Cataloging-in-Publication Data

Fernandez-Botran, Rafael.
  Methods in cellular immunology / Rafael Fernandez-Botran, Václav
Větvička.
      p.   cm.
  Includes bibliographical references and index.
  ISBN 0-8493-6548-1
  1. Cellular immunology--Laboratory manuals. I. V ětvička, Václav.
II. Title.
QR185.5.F47 1995
616.07′9′028--dc20   95-14476
                                                                                          CIP

ADA-9849

PREFACE

The main goal behind putting together a "methods book" in cellular immunology was to actually make it a useful and easy-to-use tool for investigators employing cellular immunological techniques in their research, regardless of whether or not immunology is their main area of expertise. With this idea in mind, we have striven to present state-of-the-art techniques in a step-by-step manner with coverage of background information, alternative techniques, potential limitations or pitfalls, and other useful information. In order to make the techniques easier to follow, we have provided the reader with information regarding manufacturers and commercial sources of chemicals and reagents. It should be pointed out, however, that the criterion for our mentioning a given manufacturer or commercial source for a particular reagent is simply the experience of our own laboratories, and in no way does it imply that alternative sources are not available or recommended. Finally, we have also tried to provide each chapter with a comprehensive list of references, allowing the reader to refer back to the original information and/or techniques.

Throughout this book, we have tried to provide techniques for both human and murine models. Starting with Chapter 1, we cover the isolation and preparation of different cell populations of both human and murine origin. Chapter 2 is dedicated to flow cytometry, including techniques for labeling of antibodies and fluorescent analyses of intracellular Ca^{2+} and cell cycles. In Chapter 3, we cover the preparation and culture of clones of murine and human T cells and human B cells. Chapter 4 deals with the assessment of proliferation by both isotopic and non-isotopic techniques, and Chapter 5 covers cytotoxic cell populations, including assessment of cytotoxic activity, isolation of NK cells, and generation of LAK cells. The following chapter (Chapter 6) describes bioassay and immunoassay techniques for the measurement of interleukins 1 to 15, tumor necrosis factors, and interferons from both human and murine origin. A special effort is given in this section to point out the species specificity and potential limitations of these techniques, particularly in the bioassays. Chapter 7 covers cytokine receptors and their characterization, based on ligand-binding assays and cross-linking methods. The next three chapters (Chapters 8 to 10) focus on antibodies: from the generation of polyclonal and monoclonal antibodies, the conjugation of peptides or proteins to carriers, and the detection of antibodies and antibody-producing cells (including ELISA techniques), to the purification and modification of antibodies. The last two chapters cover the preparation of buffers and media, and a list of manufacturers, complete with addresses and telephone numbers.

THE AUTHORS

Rafael Fernandez-Botran, Ph.D., is an Assistant Professor at the Department of Pathology, Division of Experimental Immunology and Immunopathology, School of Medicine, University of Louisville, Louisville, Kentucky. Dr. Fernandez-Botran received a Bachelor in Science degree in Biological Chemistry from the San Carlos University, Guatemala, in 1979; and a Ph.D. in Microbiology, from the University of Kansas Medical Center, Kansas City, KS, in 1985. Following his Ph.D. work, Dr. Fernandez-Botran moved to the Department of Microbiology at the University of Texas Southwestern Medical Center, Dallas, TX, where he worked as a post-doctoral fellow and Instructor in the laboratories of Dr. Ellen S. Vitetta and Dr. Jonathan W. Uhr. During this time, his work centered on the study of cytokines, particularly interleukin-4, and their roles in the proliferation and differentiation of CD4$^+$ T cell subsets and B cells, and on the characterization of interleukin-4 receptors and their interactions with interleukin-2 receptors. More recently, Dr. Fernandez-Botran's major research interest has been the biological role of soluble cytokine receptors. Throughout his career, Dr. Fernandez-Botran has been recipient of several fellowships and awards, including the New Investigator Recognition Award from the Clinical Immunology Society (CIS) in 1989, and the President's Young Investigator Award, from the University of Louisville in 1993. His research activities have resulted in 38 different publications, including articles, several reviews and chapters.

Václav Větvička, Ph.D., is an Assistant Professor at the Department of Pathology, Division of Experimental Immunology and Immunopathology of the School of Medicine, University of Louisville, Louisville, Kentucky. Dr. Větvička graduated in 1978 from Charles University in Prague, Czech Republic with a doctorate degree in biology and obtained his Ph.D. degree in 1983 from the Czechoslovak Academy of Sciences, Institute of Microbiology, Prague. He is a member of the Czech Immunological Society, American Association of Immunologists, and International Society of Developmental and Comparative Immunology. In 1984 he was awarded the Distinguished Young Scientist Award of the Czechoslovak Academy of Sciences. During 1984 to 1985 and 1988 he spent 18 months as a Research Associate in Professor Kincade's laboratory in Oklahoma Medical Research Foundation, Oklahoma City, Oklahoma. He is author or co-author of more than 130 original papers, several review articles, and three patents. He is a co-author of the books *Evolution of Immune Reactions* and *Immunology of Annelids* and co-editor of books *Immunological Disorders in Mice* and *Immune System Accessory Cells*, all recently published by CRC Press, Inc. Dr. Větvička's current major research interests include the role of macrophages in regulation of the immune responses, the role of complement receptors in NK cytotoxicity, and the phylogenic aspects of defense reactions.

TABLE OF CONTENTS

CHAPTER 1
Cells

ACKNOWLEDGMENT

We are indebted to Mrs. Jana Větvičkova for her valuable comments and indispensable help in the preparation of this book

DEDICATION

To Anna, Cristina, Antonio and Jana

Chapter 1

CELLS

I. PREPARATION OF CELL SUSPENSIONS

A. HEAT INACTIVATION OF FETAL CALF SERUM

Material and Reagents

Fetal calf serum (FCS)
Thermometer
Equivalent bottle
Water bath

Protocol:

1. Remove serum from -20°C freezer and allow to thaw overnight in a refrigerator.
2. Fill an equivalent bottle with water and place a thermometer in the water. Fill a water bath to 0.25 in. above the serum line.
3. Preheat the water bath to 56°C. Place the bottles with FCS and a bottle with a thermometer in the water bath.
4. Mix the serum every 10 minutes to avoid gelling of proteins.
5. When the temperature reaches 56°C, begin timing. Heat inactivate for 30 minutes with mixing continued every 10 minutes.
6. Cool rapidly by placing on ice.
7. Freeze the whole bottle or smaller aliquots.

Comments:

1. The occasional appearance of turbidity or flocculent material does not affect the quality of serum.

1

B. SPLEEN

Materials and Reagents

RPMI 1640 medium or PBS
Squeeze bottle with 70% ethanol
Scissors and forceps
Cutting board or paper towels
60 x 15-mm Petri dishes
Stainless steel screen (Thomas)
Glass tissue homogenizer (CMS)
Fetal calf serum (FCS)
15-ml conical centrifuge tube

Protocol:

1. Kill mouse either by cervical dislocation or CO_2 inhalation. Place the mouse on cutting board (or on paper towel) and soak it with ethanol to reduce the possibility of hair becoming airborne.
2. Make a cut through the skin in the inguinal region. With fingers on both sides of the cut, pull toward the tail and head until the peritoneal wall is sufficiently exposed. Soak the peritoneal wall with ethanol.
3. Cut the peritoneal wall, lift the spleen with the forceps and separate it from connecting tissues with scissors. Put the spleen into a Petri dish containing medium.
4. Cut the spleen into several smaller pieces and make a suspension.
5. Gently tease pieces of tissues over stainless steel screens using a plunger of a 5-ml syringe or alternatively use a glass homogenizer.
6. Remove large debris and cell clumps by layering the cell suspension over 3 ml of heat-inactivated FCS for 10 minutes on ice. Wash the suspension by centrifugation at 300 x g for 10 minutes at 4°C and keep in medium with at least 5% FCS.

Comments:

1. For tissue culture studies, do the whole isolation in a sterile hood and use only sterile instruments, media and dishes. Do not cut the peritoneal wall with the same scissors you used for cutting the skin.
2. Keep cells on ice.
3. The same procedure can be used for lymph nodes or thymus.
4. Typical yield per normal spleen of an untreated mouse is 5 x 10^7 to 2.5 x 10^8 cells, depending on the strain used.

2

C. BONE MARROW

Materials and Reagents

RPMI 1640 medium or PBS
Squeeze bottle with 70% ethanol
Scissors and forceps
Cutting board or paper towels
60 x 15-mm Petri dishes
3-ml syringes
22G1 needles
15-ml conical centrifuge tube

Protocol:

1. Kill mouse either by cervical dislocation or CO_2 inhalation. Place the mouse on its back on cutting board (or on paper towel) and soak it with ethanol to reduce the possibility of hair becoming airborne.
2. Make a long transverse cut through the skin in the middle of the abdominal area. Reflect skin from the hindquarters and the hind legs.
3. Separate the legs from the body at the hip joint, remove the feet. Place the legs in a Petri dish containing medium.
4. Remove all muscle tissue from the femur and tibia, separate femur and tibia and cut off the epiphyses on both ends.
5. Puncture the bone end with a needle and flush out the bone marrow with 3 ml of media.
6. Remove the large debris and cell clumps by layering the cell suspension over 3 ml of heat-inactivated FCS for 10 minutes on ice. Wash the suspension by centrifugation at 300 x g for 10 minutes at 4°C and keep in medium with at least 5% FCS.

Comments:

1. For tissue culture studies, do the whole isolation in a sterile hood and use only sterile instruments, media and dishes. Do not use the same scissors you used for cutting the skin for muscle removal.
2. Keep cells on ice.
3. Typical yield of cells is 3 to 7 x 10^7 per mouse.

3

D. PERITONEAL CELLS

Materials and Reagents

RPMI 1640 medium or PBS
Squeeze bottle with 70% ethanol
Scissors and forceps
Cutting board or paper towels
5-ml syringes
22G1 and 23G1 needles
Thioglycollate medium (Fluid Thioglycollate Medium, Becton Dickinson)
15-ml conical centrifuge tubes

Protocol:

1. Kill mouse either by cervical dislocation or CO_2 inhalation. Place the mouse on its back on cutting board (or on paper towel) and soak it with ethanol to reduce the possibility of hair becoming airborne.
2. Make a cut through the skin in the inguinal region. With fingers on both sides of the cut, pull toward the tail and head until the peritoneal wall is sufficiently exposed. Soak the peritoneal wall with ethanol.
3. Inject 5 ml of medium into the peritoneal cavity using a 22G1 needle. Gently massage the peritoneum and slowly draw the fluid using a 23G1 needle. You may need to puncture the cavity in several places. Wash the suspension by centrifugation at 300 x g for 10 minutes at 4°C and keep in medium with at least 5% FCS.

Comments:

1. The average yield is 2 x 10^6 cells per mouse. Approximately 30 to 40% of cells are macrophages.
2. You can substantially increase the yield of cells (percentage of macrophages will be more than 90%) by injection of 3 ml of thioglycollate medium i.p. 3 to 6 days prior to harvesting peritoneal cells. The yield depends on the strain, being between 2 x 10^7 per mouse (A/J or BALB/c) and 2 to 3 x 10^7 (C57BL/6, C57BL/10).[1]
3. If you use thioglycollate medium, remember three important things: first, the macrophages are activated and thus you can not compare them with macrophages from untreated mice;[1,2] second, always use the same interval between thioglycollate injection and cell harvesting; and third,

thioglycollate medium is a bacteriological medium, therefore you must keep it sterile.

E. CELL COUNTING

Materials and Reagents

Cell suspension
Hemocytometer (CMS)
Pipettes
Pasteur pipettes
Eppendorf tubes
Türk or Trypan blue solution

Protocol:

1. Dilute cells in either Türk or Trypan blue solution, mix thoroughly and add one drop into hemocytometer chamber using a Pasteur pipette. Do not overfill or underfill the chamber.
2. Count the cells in the 1-mm center square. Count cells touching the top and left middle lines of the perimeter of each square.
3. Cell counts/ml = Actual number of cells x dilution factor x 10^4

Comments:

1. It is possible to increase the accuracy by counting more than one 1-mm square (5 to 10 squares). In that case, substitute *Actual number of cells* with *Average number of cells per square.*
2. The optimal cell dilution is approximately 20 to 50 cells per square.

F. CELL COUNTING USING TÜRK SOLUTION

Materials and Reagents

Cell suspension
Türk solution
Hemocytometer
Pipettes
Eppendorf tubes

Protocol:

1. Dilute cells in Türk solution, mix thoroughly and add one drop into hemocytometer chamber using a Pasteur pipette and count them.

Comments:

1. Türk solution lyses all erythrocytes, so it is necessary to use it for counting of either spleen cells or cell suspension resulting from isolation from peripheral blood.

G. DETERMINATION OF CELL VIABILITY

1. Trypan Blue Exclusion

Materials and Reagents

Cell suspension at 2 to 6 x 10^6 cells/ml
Trypan blue (Sigma), 0.4% (w/v) in water
PBS
5-ml glass or plastic tube

Protocol:

1. Mix 0.25 ml of trypan blue solution, 0.15 ml of PBS and 0.1 ml of the cell suspension. Allow to stand for 5 to 15 minutes.
2. Transfer a small amount of the suspension either to the hemocytometer chamber or on cover glass and count the cells. Nonviable cells will stain blue.

$$\text{Cell viability (\%)} = \frac{\text{number of viable cells}}{\text{number of viable cells} + \text{number of dead cells}} \times 100$$

Comments:

1. Trypan blue has a greater affinity for serum proteins than for cellular proteins. If the background is too dark, resuspend the cells in PBS prior to counting.

2. Do not incubate cells in trypan blue solution longer than 15 minutes, as viable cells might begin to take up dye.

2. Fluorescein Diacetate

Materials and Reagents

Cell suspension at approximately 2×10^6 cells/ml
Fluorescein diacetate (Sigma), 5 mg/ml in acetone, stored at -20°C
PBS
FCS
Fluorescence microscope or flow cytometer
15-ml conical centrifuge tube

Protocol:

1. Dilute fluorescein diacetate in PBS at room temperature (1:50). Add 0.1 ml of diluted fluorescein diacetate to 0.9 ml of cell suspension.
2. Incubate for 5 to 10 minutes at room temperature. Examine cells with a fluorescence microscope or with a flow cytometer. Viable cells will appear bright green, dead cells will not be stained at all. Determine the percent of viable cells as described above.

Comments:

1. Fluorescein diacetate is taken in by all cells but only shows green fluorescence after hydrolyzation inside living cells.
2. This technique is particularly useful in combination with rhodamine-(TRITC) or phycoerythrin-labeled antibodies.

2. Propidium Iodide

Materials and Reagents

Cell suspension
Propidium iodide (Sigma), 1 mg/ml in PBS
5-ml centrifuge tubes or 1.5-ml Eppendorf tubes
Fluorescence microscope or flow cytometer

Protocol:

1. Mix 200 µl of cells with 10 µl of propidium iodide and incubate for 5 minutes at 4°C.
2. Fill the tube with washing medium and wash twice by spinning 10 minutes at 200 x g.
3. Examine cells with a fluorescence microscope or with a flow cytometer. Dead cells will appear red, viable cells will not be stained at all. Determine the percent of viable cells as described above.

Comments:

1. This technique can be used in combination with fluorescein diacetate.
2. Use propidium iodide staining at time of last incubation in every cell-surface staining experiment using FITC-labeled antibodies and flow cytometer.
3. Do not use in combination with TRITC- or phycoerythrin-labeled antibodies.

H. REMOVAL OF ERYTHROCYTES FROM CELL SUSPENSIONS

1. ACK Lysing Buffer

Materials and Reagents

Cell suspension
ACK lysing buffer
15-ml conical centrifuge tubes

Protocol:

1. Add spleen cell suspension into the lysing buffer. Use approximately 5 ml per spleen.
2. Incubate for 5 minutes at room temperature with occasional gentle shaking.
3. Fill the tube with washing medium (optimally with medium you plan to use in subsequent experiment), wash twice by centrifugation for 10 minutes at 200 x g.

2. Hemolytic Gey's Solution

Materials and Reagents

Cell suspension
Gey's solution
FCS
15-ml conical centrifuge tubes

Protocol:

1. Add 5 ml of Gey's solution to 1 x 10^8 spleen cells and incubate for 5 minutes on ice.
2. Underlay the cells with 100% FCS and spin at 300 x g for 10 minutes at room temperature. Wash twice with washing medium by spinning at 300 x g for 10 minutes at room temperature.

3. Tris-Buffered Ammonium Chloride

Materials and Reagents

Cell suspension
FCS
Tris-buffered ammonium chloride
15-ml conical centrifuge tube

Protocol:

1. Resuspend 1 x 10^8 cells in 3 ml of Tris-buffered ammonium chloride (working solution) and incubate for 2 minutes at room temperature.
2. Underlay the cells with 100% FCS and spin at 300 x g for 10 minutes. Wash twice with washing medium.

Comments:

1. Regardless of which method you use, if the cell suspension still looks red, it is necessary to repeat the whole process.
2. The removal of red blood cells is necessary for spleen cell suspension only.
3. Remove red blood cells before counting the cells.

I. REMOVAL OF DEAD CELLS

1. Centrifugation Through FCS

Materials and Reagents

Cell suspension
FCS
15-ml conical centrifuge tube

Protocol:

1. Carefully layer 1×10^8 cells in 1 ml of medium over 3 ml FCS in a 15-ml conical centrifuge tube. Centrifuge at 300 x g for 10 minutes.
2. Discard the supernatant and wash the cells twice in an appropriate medium.

2. Centrifugation Through Ficoll-Hypaque™

Materials and Reagents

Cell suspension
Ficoll-Hypaque™ solution ϱ = 1.09 (Sigma; Hypaque-76 from Winthrop Pharmaceuticals)
RMPI 1640 medium with 5% FCS
15-ml conical centrifuge tube

Protocol[3]:

1. Add 4 ml of Ficoll-Hypaque™ solution to the 15-ml conical tube and carefully layer 4 ml of cells (5×10^6 to 1×10^7/ml) over Ficoll-Hypaque™ (use the wall of the tube).
2. Spin at 2,000 x g for 20 minutes at room temperature with brake off.
3. Collect all of the fluid, add medium with FCS and centrifuge at 300 x g for 15 minutes at 4°C. Wash twice in an appropriate medium at 250 x g for 10 minutes.

Comments:

1   The density of Ficoll-Hypaque™ mixture is temperature dependent. It is necessary to use the prewarmed Ficoll-Hypaque™ (20°C) and centrifuge.

II. ISOLATION OF CELLS

A. ISOLATION OF B CELLS

1. B Cell Enrichment by Cytotoxic Elimination of T Cells

Materials and Reagents

Cell suspension
Antibodies of known specificity and cytotoxic titer
RPMI 1640 medium with 10% FCS
Rabbit complement
15-ml conical centrifuge tubes

Protocol:

1. Dilute cell suspension to 10^7 cells/ml. Add 1 ml of antibodies and incubate for 30 minutes.
2. Wash the cell suspension in medium at 250 x g for 10 minutes at room temperature.
3. Add 1 ml of diluted rabbit complement and incubate for 30 minutes at 37°C.
4. Wash the cell suspension three times in medium at 250 x g for 10 minutes at room temperature.

Comments:

1. The optimal antibodies are anti-Thy-1 (in the case of mouse cells) or anti-CD3 (in the case of human cells). The use of either anti-CD4 or anti-CD8 antibodies will result in selective elimination of either CD4$^+$ or CD8$^+$ T cells.[4]

2. Panning[5]

Materials and Reagents

Cell suspension
Anti-Ig antibody (or anti-CD19, anti-CD20), affinity purified

11

PBS
15 x 100-mm Petri dishes
RPMI 1640 medium with 5% FCS

Protocol:

1. Coat the Petri dish with 10 ml of antibody diluted to 50 to 100 µg/ml. Be sure that the whole surface of the dish is covered. Incubate for 2 hours at room temperature or overnight at 4°C.
2. Remove the anti-Ig solution and wash the plate 4x with cold PBS.
3. Add cells (up to 2×10^8) in 5 ml of medium with FCS and incubate for 30 minutes at room temperature.
4. Remove nonadherent cells (T cells).
5. Pour 10 ml of medium with FCS.
6. With a pipette or syringe draw most of the medium off and strongly pour the medium back into one area of the dish. Repeat several times over a new area until the whole dish is covered and most of the cells are removed.

Comments:

1. The coated plates can be stored in the refrigerator for up to 4 weeks.
2. The anti-Ig solution can be stored and used repeatedly.
3. This method can also be used for isolation of T cells.
4. Check the efficiency of isolation by staining the final suspension with anti-Ig (anti-CD19/20) or anti-CD3 antibodies conjugated to a fluorescent marker.

3. Isolation of B Cells Using Percoll Gradient

<u>Materials and Reagents</u>

Hanks' balanced salt solution (HBSS)
Percoll (Pharmacia LKB; Sigma)
50-ml polypropylene conical centrifuge tubes
15-ml glass centrifuge tubes
Cell suspension

Protocol:

1. Prepare 70% Percoll by mixing 170 ml Percoll mix solution with 290 ml Percoll.

12

2. Prepare 50%, 60% and 66% Percoll as follows:

	50%	60%	66%
HBSS (ml)	8.58	4.26	1.71
70% Percoll (ml)	21.42	25.74	28.29

3. Prepare a 50%-60%-66%-70% Percoll gradient in 15-ml tubes. Start with 2.5 ml of 70% Percoll, gently overlay 2.5 ml of 66% Percoll followed by 2.5 ml of 60% Percoll and finally 2.5 ml of 50% Percoll. Incubate the gradients for 15 minutes on ice.
4. Gently add 2.5 ml of cell suspension (approximately 5×10^7/ml) on the top of the gradient.
5. Centrifuge gradients for 13 minutes at 1,900 x g at 4°C.
6. Aspirate HBSS and 50% Percoll and separately collect cells at interfaces between remaining layers.
7. Wash cells three times with HBSS by centrifugation at 300 x g at 4°C.

Comments:

1. Best results are achieved when this method is used as a supplement to any of the previously described methods for enrichment of B cell populations.
2. Keep HBSS and all Percoll solutions on ice all the time, the success of the separation depends on the temperature.
3. Cells at the 66%-60% Percoll interface are resting B lymphocytes, cells at the 50%-60% Percoll interface are activated B cells.
4. If a sterile cell suspension is needed, perform all steps in a laminar flow hood.

B. ISOLATION OF T CELLS

1. Adherence to Nylon Wool[6]

Materials and Reagents

Cell suspension
RPMI 1640 medium with 5% FCS
50-ml conical centrifuge tubes

13

Nylon wool (Polysciences)
10-, 30- or 60-ml disposable syringes
3-way stopcock
HCl (1%)
Glass beaker

Protocol:

1. Boil the nylon wool in 1% HCl in a glass beaker for 20 minutes. Wait until the fluid is cold and then pour it off. Wash the nylon wool repeatedly to remove all HCl and check the pH. Let the nylon wool dry at room temperature, and weigh in appropriate aliquots. You can store the dry nylon wool indefinitely.

2. Prepare the nylon wool column by packing a syringe of an appropriate size with preboiled nylon wool. Two grams of nylon wool is suitable for separation of approximately 3×10^8 cells in 4 ml.

3. Equilibrate the column by running 10x the loading volume of warm (37°C) medium. Remove all trapped air bubbles by tapping on the column sides. Compact the nylon wool with a pipette.

4. Close the 3-way stopcock and add an extra 2 to 3 ml of warm medium. Incubate the column for 30 minutes at 37°C.

5. Drain the column completely and add the prewarmed cell suspension (approximately 7.5×10^7 cell/ml).

6. Drain the column, add 1 to 2 ml of warm medium and drain again. Add 3 ml of media to prevent drying and close the stopcock.

7. Incubate the column for another 30 minutes at 37°C.

8. Fill the column with warm medium, open the stopcock and collect the first 15 to 20 ml of nonadherent cells.

9. Wash the cell suspension twice in an appropriate medium at 250 x g for 10 minutes at 4°C.

Comments:

1. If you need a sterile cell suspension, perform all steps in a laminar flow hood.

2. Do not overload the column.

3. For further use of nylon wool, repeat step # 1.

2. T Cell Enrichment by Cytotoxic Elimination

Materials and Reagents

Cell suspension
Antibodies of known specificity and cytotoxic titer
RPMI 1640 medium with 5% FCS
Rabbit complement
15-ml conical centrifuge tubes

Protocol:

1. Dilute cell suspension to 10^7 cells/ml. Add 1 ml of antibodies and incubate for 30 minutes.
2. Wash the cell suspension in medium at 250 x g for 10 minutes at 4°C.
3. Add 1 ml of diluted rabbit complement and incubate for 30 minutes at 37°C.
4. Wash the cell suspension three times in medium at 250 x g for 10 minutes at 4°C.

Comments:

1. Several pilot experiments are crucial prior to the actual separation. The optimal conditions vary according to the antibody, cells and complement used. After establishing the optimal conditions, controls (addition of antibodies without complement and addition of complement without antibodies) need not be routinely performed.
2. The antibodies used depend on the type of cells you are eliminating. Anti-MHC class II antibodies and/or anti-CD19 or CD20 antibodies work fine for elimination of B cells, anti-CD14 antibodies for elimination of monocytes, anti-F4/80 antibodies for elimination of mouse macrophages. The use of purified IgG is not necessary; culture supernatants or ascites fluids work just fine. The optimal concentration must be determined in pilot experiments. The temperature for incubation also varies according to the antibodies used and thus must be determined in pilot experiments.
3. Test the new lot of rabbit complement for spontaneous cytotoxicity (without addition of antibodies). If possible, prepare your own complement (by bleeding the rabbit, incubate serum for 20 minutes at room temperature followed by 120 minutes at 4°C). If you decide to use commercial rabbit complement, the best results are achieved with complement from Cedarlane. After determination of ideal dilution, store

aliquots of complement at -80°C for not longer than 1 year. Dilute the complement just before use and always discard the unused part.

4. The temperature for incubation of cells with antibodies varies also according to the antibody used. We would recommend performing the first test with 4°C and moving to either room temperature or even 37°C only if the results are not satisfactory.

C. ISOLATION OF MONOCYTES/MACROPHAGES

1. Adherence to Sephadex G-10[7]

Materials and Reagents

Cell suspension
RPMI 1640 medium with 5% FCS
Sephadex G10 (Pharmacia; Sigma)
10- or 50-ml disposable syringe
3-way stopcock
50-ml conical centrifuge tube
Nylon wool (Polysciences)
1% SDS in PBS
Pasteur pipette

Protocol:

1. Prepare an empty syringe with a closed 3-way stopcock and a needle and prewarm RPMI 1640 medium.
2. Put approximately 10 mg of nylon wool into the bottom of the syringe and press well with the tip of a pipette. Add 5 to 10 ml of medium with FCS and remove all air bubles. With the tip of a pipette or Pasteur pipette hold the nylon wool firmly down and open the stopcock to seal the syringe. Close the stopcock.
3. Mix Sephadex G-10 with warm (37°C) medium with FCS.
4. Add 5 to 10 ml of Sephadex G-10 into the syringe and allow to settle down. Open the stopcock, but do not allow all the medium to drain from the Sephadex G-10 mixture. Add the rest of the Sephadex G-10 (leave space for approximately 3 ml of cells). Wash the column with 50 to 150 ml of warm medium with FCS. Close the stopcock.
5. Add the cell suspension (1×10^8) dropwise onto the column, open the stopcock and allow the cells to penetrate the column. Close the stopcock and incubate for 30 minutes at 37°C.
6. Open the stopcock and collect the nonadherent cells. Add additional

16

warm medium, but collect only about 10 to 20 ml (depending on the size of the column).

7. Let the possible contaminating Sephadex G-10 to settle in a tube (1 to 2 minutes), decant the medium into a new tube, and wash the cell suspension in medium at 250 x g for 10 minutes at 4°C.

Comments:

1. If you need a sterile cell suspension, perform all steps in a laminar flow hood.
2. In order to completely deplete the monocytes/macrophages, repeat the whole process one more time.
3. Collect the used Sephadex G-10. Wash repeatedly with an excess of water. Pour off the water, add a 1% solution of sodium dodecyl sulfate in PBS and incubate at room temperature for at least 12 hours. Pour off the SDS solution and wash repeatedly with an excess of water (at least 10x). Wash 5x with PBS. Resuspend Sephadex G10 in a small volume of PBS and store at 4°C. Sephadex G10 can be used up to five times.

2. Isolation of Monocytes by Adhesion and Cultivation

Material and Reagents

Peripheral blood
3.8 % sodium citrate
24-well tissue culture plates
15-ml and 50-ml conical centrifuge tubes
Ficoll-Histopaque (Sigma)
1 M CaCl$_2$
6% Dextran T-500 (Sigma)
Saline
16 x 125-mm tubes
Hanks' buffered saline
RMPI 1640 medium
Antibiotics
Trypsin-EDTA solution (BioWhittaker; Intergen)

17

Protocol:

1. Centrifuge freshly drawn peripheral blood with 0.11 ml of 3.8% sodium citrate/ml of blood at 1,000 x g at room temperature for 20 minutes.
2. Remove plasma from cells. Place approximately 8 ml of plasma into new tube and spin for 15 minutes at 1,000 x g at room temperature.
3. Add 5 ml of 6% dextran to cells. Bring volume up to 50 ml with saline. Invert several times, loosen cap and let sit for 30 minutes at room temperature.
4. Put the remaining plasma into a sterile bottle and add 20 μl $CaCl_2$/ml of plasma (this is now considered to be an autologous serum). Incubate at 37°C for 1 hour.
5. Dilute part of the serum 3x with saline.
6. Remove supernatant from the dextran sedimentation tube, spin at 350 x g for 10 minutes at room temperature. Discard supernatant and save cells.
7. Resuspend these cells in diluted serum (1 to 4 x 10^6/ml). Place 8 ml of cell solution into each 16 x 125 mm tube and underlay with 3 ml of Ficoll-Histopaque. Centrifuge at 1,000 x g for 25 minutes at room temperature.
8. Monocytes are in the band at the interface of the gradient. Remove this layer, put the cells into 50-ml tube and fill with cold Hanks' buffered saline. Wash twice by centrifugation at 350 x g for 10 minutes at 4°C.
9. Dilute cells to 2 to 3 x 10^6/ml with Hanks' buffered saline. Add 0.1% autologous serum (from step # 4) and place 1 ml of cells/well in a 24-well tissue culture plate. Incubate for 2 hours at 37°C.
10. Aspirate nonadherent cells, wash once with RPMI 1640, add 2 ml of RPMI 1640 medium supplemented with 5% autologous serum and antibiotics and incubate for 3 to 4 days in a humidified 37°C, 5% CO_2 incubator.
11. Resulting monocytes/macrophages can be removed by Trypsin-EDTA solution (aspirate medium from wells, add 0.2 ml of Trypsin-EDTA solution into each well and incubate for 5 to 10 minutes at 37°C. Check under the microscope. Pool cells from all wells into 50-ml tube, add RMPI 1640 medium and centrifuge at 350 x g for 10 minutes at 4°C).

Comments:

1. The resulting population depends on the time of incubation. It is either a mixture of monocytes and macrophages (with a minor contamination of other cell types), or in the case of 4-day or longer incubations, it consists of macrophages only.

18

2. If you need a sterile cell suspension, perform all steps in a laminar flow hood.

3. Isolation of Monocytes by Adherence

Materials and Reagents

Peripheral blood mononuclear cells
RMPI 1640 medium
75-cm^2 tissue culture flasks (Corning; Costar)
15-ml conical centrifuge tube
Trypsin-EDTA (BioWhittaker; Intergen) or rubber policeman (Costar)

Protocol:

1. Dilute cells in RMPI 1640 medium to a concentration 2 x 10^6/ml.
2. Put 10 ml of diluted cells into 75-cm^2 tissue culture flask and incubate 60 minutes in a humidified 37°C incubator.
3. Aspirate and discard the medium, wash flasks twice with 10 ml of RPMI 1640 medium and remove adherent cells either by rubber policeman or Trypsin-EDTA treatment.

Comments:

1. Some authors recommend using medium supplemented with 10% FCS, human serum or autologous human serum.[8]
2. If you need a sterile cell suspension, perform all steps in a laminar flow hood.

4. Removal or Purification of Monocytes[9]

Materials and Reagents

Cell suspension
Baby hamster kidney (BHK) cells (ATCC # 6281)
RPMI 1640 medium with 10% FCS
RPMI 1640 medium with 1% FCS
PBS
75-cm^2 tissue culture flasks (Corning; Costar)

19

10 mM and 3.3 mM EDTA in PBS
Rubber policeman (Corning; Costar)

Protocol:

1. Grow baby hamster kidney cells to confluency in RPMI 1640 medium supplemented with 10% FCS.
2. Remove them from the culture flask with 10 mM EDTA-PBS and rinse each flask three times with PBS.
3. Incubate 2×10^7 cells (the cell suspension you want to purify) in 10 ml of RPMI 1640 medium with 1% FCS in the flask where the BHK cells were cultured. Incubate at 37°C for 60 minutes.
4. Decant the nonadherent cells (i.e., population depleted of monocytes) and rinse the flask three times with PBS. Adherent monocytes are removed after a 10 minute incubation with 3.3 mM EDTA at 37°C by gentle scraping with a rubber policeman.

Comments:

1. The baby hamster kidney cell-pretreated tissue culture flasks can be stored in a -30°C freezer for up to 6 months.

5. Removal of Monocytes/Macrophages Using L-Leucine Methyl Ester[10]

Materials and Reagents

Cell suspension
1 mM L-leucine methyl ester (Sigma)
RPMI 1640 medium
RPMI 1640 medium with 10% FCS
FCS
15-ml conical centrifuge tubes

Protocol:

1. Incubate 1×10^7 cells in serum-free RPMI 1640 medium containing 1 mM L-leucine methyl ester. Incubate for 40 minutes at room temperature.
2. Wash twice in serum-free RPMI 1640 medium by centrifugation at 300 x g for 10 minutes at 4°C and resuspend in RPMI 1640 medium with 10% FCS.

Comments:

1. NK cells and cytotoxic T cells will be also killed.

D. PURIFICATION OF CELL POPULATIONS USING MAGNETIC MICROSPHERES

Materials and Reagents

Magnetic microspheres (Dynal Inc.; Miltenyi Biotec Inc.; Advanced Magnetics)
Cell suspension
Appropriate monoclonal antibodies (in case microspheres are not directly coupled to antibodies)
Centrifuge tubes (size according to the type of magnetic device)
RPMI 1640 medium
Plastic Pasteur pipettes
Rotator
Magnet

Protocol:

1. Incubate cells with monoclonal antibodies in centrifuge tube for 30 minutes at 4°C on a rotator with end-over-end rotation. The most common concentration is 1 µg/10^6 cells, but pilot flow cytometry experiments might allow use of even lower concentrations.
2. Wash cells twice by centrifugation in RPMI 1640 medium at 150 x g at 4°C.
3. Add anti-Ig-coated magnetic microspheres according to manufacturer's instructions and incubate for 60 minutes at 4°C on a rotator with end-over-end rotation.
4. Separate microsphere-coated cells using the magnetic apparatus. After 5 minutes, carefully transfer the unbound cells to a new tube using a plastic Pasteur pipette and repeat the magnetic separation.
5. Wash cells twice by centrifugation in RPMI 1640 medium at 150 x g at 4°C and count them.
6. Before the technique becomes routine (and after using a new batch of antibodies), analyze the isolated cell population by flow cytometry.

Comments:

1. This technique can be used for depletion (negative selection) or isolation (positive selection) of any cell subset against which appropriate monoclonal antibodies are available. Thus, we will not repeat this method in individual sections of this chapter.

2. You can combine several monoclonal antibodies in one step, e.g., use mouse anti-CD14, mouse anti-CD19 and mouse anti-CD8 and subsequently anti-mouse Ig-coated magnetic microspheres for depletion of monocytes, B cells and CD8$^+$ T cells. The separation will result in pure CD4$^+$ T cells.

3. For best results, repeat isolation steps twice.

4. If you need a sterile cell suspension, perform all steps in a laminar flow hood.

5. Even if optimal results are achieved using magnetic devices available from a company supplying immunobeads, it is not necessary to purchase several rather expensive types of magnets, as the use of different magnets is only less convenient.

6. Use of directly labeled magnetic beads allows you to skip steps # 1 and 2 and speeds up the whole procedure.

7. Several studies prefer diferent types of microspheres. However, in our hands, the best results were achieved using the Dynal beads. Microspheres from Miltenyi Biotec Inc. are so small that for most experiments (such as flow cytometry) you do not need to remove the beads from the cells, which substantially speeds up the isolation.

8. A modification of this technique is the use of new DETACHaBEAD from Dynal, Inc. This product is developed for rapid detachment of cells bound to Dynabeads. DETACHaBEAD is a special polyclonal antibody preparation that reacts with the F(ab) fragments of mouse monoclonal antibodies, subsequently disturbing and blocking the interaction between Dynabeads and the cell surface. The whole technique represents a double magnetic isolation and results in a purified population of the selected cell type (i.e., positive separation). The only disadvantage so far is the limited number of Dynabeads available (the new Dynabeads specifically designed for use with DETACHaBEAD are coated with CD4, CD8 and CD19 only), but the manufacturer promises to develop the whole range of Dynabeads.

E. PURIFICATION OF CELLS USING PREPARED COLUMNS

Materials and Reagents

RPMI 1640 medium with 5% FCS
Pipettes
Separation Cellect column (Biotex)
Centrifuge tubes
Cell suspension

Protocol:

1. Reconstitute column reagents (provided).
2. Add these reagents to a column and incubate 60 minutes at room temperature.
3. Add RPMI 1640 medium with 5% FCS and set flow rate according to the manufacturer's instructions.
4. Add cell suspension and incubate for 30 minutes at room temperature.
5. Add RPMI 1640 medium with 5% FCS and collect enriched cells.

Comments:

1. The number of cells per column is different according to the type of the column used.
2. At the present time, columns prepared for isolation of human, mouse and rat T lymphocytes, CD4 and CD8 lymphocytes are available.

F. ISOLATION OF PERIPHERAL BLOOD CELLS

1. Ficoll-Hypaque Separation of Neutrophils and Mononuclear Cells

Materials and Reagents

Peripheral blood
3.8% sodium citrate in H_2O
50-ml conical centrifuge tubes
3% Dextran T-500 (Sigma)
PBS
Ring stand
30-ml disposable syringes
18G1 needle

Tygon tubing 3/32 x 5/32 x 1/32 (CMS)
Ficoll-Hypaque/Histopaque
Plastic transfer pipettes (CMS)
PE-160 Intramedic polyethylene tubing

Protocol:

1. Draw desired amount of fresh peripheral blood into a syringe containing sodium citrate (5 ml per 60 ml blood).
2. Mix the blood with an equal volume of 3% Dextran T-500 in PBS.
3. Draw the mixture into syringes, and place the syringe on a ring stand, with the tip facing up. Place Tygon tubing on the tip.
4. Incubate 30 min at room temperature to allow sedimentation of red blood cells.
5. Place the leukocyte-rich plasma into 50-ml centrifuge tubes, 10 to 12 ml per tube.
6. Underlayer the mixture with 12 ml of 1.08 g/ml Ficoll-Hypaque using a 30-ml syringe with 18-gauge needle and PE-160 Intramedic polyethylene tubing.
7. Then underlayer the first layer with 12 ml of 1.105 g/ml Ficoll-Hypaque.
8. Centrifuge for 30 min at 1,000 x g at room temperature.
9. Harvest the cells with a plastic transfer pipette from the two interfaces from each tube: the top white cell layer between the plasma and the 1.08 Ficoll-Hypaque is a mixture of mononuclear cells (lymphocytes and monocytes) and platelets; the bottom layer between the 1.08 and 1.105 Ficoll-Hypaque contains mostly neutrophils.
10. Pool the homologous layers from up to 3 tubes into a 50-ml conical tube and dilute to full tube capacity with PBS (room temperature).
11. Centrifuge at 400 x g for 12 min at room temperature.
12. Decant and blot the tubes.
13. Resuspend and wash cells at least two times with PBS, centrifuging at 250 x g for 10 min.
14. Resuspend the cells in the desired media and count.

2. Isolation of Mononuclear Cells Using Sepracell-MN

Materials and Reagents

Peripheral blood
Sepracell-MN (Sepratech Corp.)

24

PBS-0.1% BSA (included with Sepracell-MN)
50-ml conical centrifuge tubes
3.8% sodium citrate or 0.1 M EDTA
Plastic transfer pipettes

Protocol:

1. Draw desired amount of fresh peripheral blood into a syringe containing 3.8% sodium citrate or 0.1 M EDTA.
2. Add equal volumes of well-mixed blood and Sepracell-MN to a centrifuge tube, tighten cap and mix gently by inversion several times.
3. Centifuge at 1,500 x g at room temperature for 20 minutes.
4. Harvest cells (lymphocytes, monocytes and platelets) with a plastic transfer pipette from the band just below the meniscus.
5. Mix the cells with four volumes of PBS and wash once at 300 x g for 10 minutes.
6. Resuspend in the desired media and count.

Comments:

1. Use 20% less volume of Sepracell-MN with 24 hour or older EDTA-blood samples.
2. The same procedure can be used for isolation of mononuclear cells from both human and mouse blood.

3. Isolation of Monocytes and Lymphocytes Using Sepracell-MN

<u>**Materials and Reagents**</u>

Peripheral blood
Sepracell-MN (Sepratech Corp.)
PBS-0.1% BSA (included with Sepracell-MN)
50-ml conical centrifuge tubes
3.8% sodium citrate or 0.1 M EDTA
Plastic transfer pipettes

Protocol:

1. Wash cells only once following separation of mononuclear cells with Sepracell-MN (see Section 2).

25

2. Mix cells and Sepracell-MN in a ratio of 2:1 in a centrifuge tube, tighten cap and mix gently by inversion several times.
3. Centifuge at 1500 x g at room temperature for 20 minutes.
4. Collect monocytes (and platelets) from the band just below the meniscus and/or lymphocytes from a band near the bottom of the tube.
5. Mix the cells with four volumes of PBS and wash once at 300 x g for 10 minutes.
6. Resuspend in the desired media and count.

4. Isolation of Basophils

Materials and Reagents

Peripheral blood
3.8% Sodium citrate in H_2O
50-ml conical centrifuge tubes
30-ml disposable syringes
18G1 needle
Percoll
Plastic transfer pipettes
HBSS without Ca^{2+} and Mg^{2+}
0.25 M HEPES
1 M HCl

Protocol:

1. Prepare a stock solution by mixing the following:

   90 ml Percoll
   9 ml of 10x HBSS
   1 ml 0.25 M HEPES
   0.4 ml 1 M HCl

2. Prepare the following densities of Percoll:

   1.070 g/ml: 24.0 ml of Percoll stock mixed with 20.0 ml of 1x HBSS
   1.079 g/ml: 27.0 ml of Percoll stock mixed with 15.9 ml of 1x HBSS
   1.088 g/ml: 23.0 ml of Percoll stock mixed with 10.0 ml of 1x HBSS
3. Form a Percoll gradient by sequentially adding to the bottom of the tube 4 ml of 1.070 g/ml solution, 4 ml of 1.079 g/ml solution and 3 ml of 1.088 g/ml solution.

26

4. Carefully layer 4 ml of freshly collected peripheral blood over Percoll gradients and centrifuge 25 minutes at 300 x g at room temperature with brake off.
5. Collect cells from each cell band with a plastic transfer pipette. Pool same bands from each gradient tube into one 15-ml centrifuge tube. Place 3.3 ml aliquots in new, separate 15-ml tubes.
6. Wash each aliquot twice with 10 ml of HBSS by centrifugation at 300 x g at 4°C for 10 minutes.

Comments:

1. Basophils are usually found in the middle band, but due to the donor to donor variations, check all bands by Giemsa-Wright staining.
2. Peripheral blood basophils are very rare; therefore, this procedure results in 15 to 50% enrichment only (depending on the donor). You can further purify the population using some of the methods mentioned above (such as negative selection of contaminating cells by magnetic microspheres).

REFERENCES

1. **Vetvicka, V., Zeleny, V. and Viklicky, V.,** Cell surface changes of the mouse peritoneal macrophages after proteose peptone or thioglycollate induction, *Fol. Biol.,* 25, 156, 1979.
2. **Fornusek, L., Vetvicka, V. and Kopecek, J.,** Differences in phagocytic activity of methacrylate copolymer particles in normal and stimulated macrophages, *Experientia,* 37, 418, 1981.
3. **Parish, C. R., Kirov, S. M., Bowern, N., Blanden, N. and Blanden, R.V.,** A one-step procedure for separating mouse T and B lymphocytes, *Eur. J. Immunol.,* 4, 808, 1974.
4. **Swain, S. L.,** Significance of Lyt phenotypes: Lyt2 antibodies block activation of T cells that recognize class I MHC antigens regardless of their function, *Proc. Natl. Acad. Sci. U.S.A.,* 78, 7101, 1981.
5. **Wysocki, W. L. and Sato, V. L.,** "Panning" for lymphocytes: A method for cell selection. *Proc. Natl. Acad. Sci. U.S.A.,* 75, 2844, 1978.
6. **Julius, M. H., Simpson, E. and Herzenberg, L. A.,** A rapid method for the isolation of functional thymus-derived murine lymphocytes, *Eur. J. Immunol.,* 3, 645, 1973.
7. **Ly, I. A., Mishell, R. I.,** Separation of mouse spleen cells by passage through columns of Sephadex G-10, *J. Immunol. Methods,* 5, 239, 1974.

27

8. Edelson, P. J. and Cohn, Z. A., Purification and cultivation of monocytes and macrophages, in *In vitro Methods in Cell-mediated and Tumor Immunity*, Bloom, B. A. and Davis, J. R., Eds., Academic Press, San Diego, 1976, 333.

9. Ackerman, S. K. and Douglas, S. D., Purification of human monocytes on microexudate-coated surfaces, *J. Immunol.*, 120, 1372, 1978.

10. Thiele, D. L., Kurosaka, M. and Lipsky, P. E., Phenotype of the accessory cell necessary for mitogen-stimulated T and B cell responses in human peripheral blood: Delineation by its sensitivity to the lysosomotropic agent, L-leucine methyl ester, *J. Immunol.*, 131, 2282, 1983.

Chapter 2

FLOW CYTOMETRY

I. OVERVIEW

The principle of flow cytometry is rather simple: to scan single cells flowing past an excitation source in a liquid medium and analyze them according to their fluorescence. In a typical flow cytometer, a suspension of fluorescently labeled cells is made to flow in a laminar manner by a suitable hydrodynamic system, such that they line up in a single file. As they move, they cross a beam of light from a laser. As a result, two types of light, scattered light and fluorescent light, are transmitted.

In order to describe the various techniques used in either cell surface analysis or cell sorting, it would be beneficial to introduce some terminology. The purpose of this chapter is to survey the most commonly used techniques and to either suggest some modifications or to point out some possible pitfalls. To discuss the advantages of one cell cytometer against another or which cell cytometer would be better for your laboratory falls out of the scope of this chapter.

Flow cytometers can have one or more **light sources**. The most commonly utilized light source is a laser, since it exhibits the highest attainable spectral radiance. Readers seeking more details should consult reference 1. Each light beam is operated at a single wavelength, which will restrict the use of the various probes (see Table 1, Figure 1). Similarly, fluorescence emissions from different probes which excite at a single wavelength will often have spectral overlap (Table 1). Proper understanding of these parameters is essential for determination of which fluorescent probes we might employ. A second excitation source significantly expands the availability of probes to be used simultaneously.

The spectral overlap can be compensated using **electronic compensation**. Electronic compensation means substraction of a percentage of the signal measured from the first probe from that of the second probe. The commonly used fluorochromes, fluorescein isothiocyanate (FITC) and phycoerythrin (PE), can serve as an example. FITC is measured at a peak of 530 nm as green fluorescence, and PE at a peak of 575 nm as orange fluorescence (see Figures 2 to 4). However, a significant part of green fluorescence appears also in orange emission (and much less of orange in green emission). It is optimal to set the electronic compensation before running the real samples. The easiest way is to mix cells labeled with a single fluorochrome and unlabeled cells, run these cells and then set the electronic compensation in such a way that the amount of orange (in the case of FITC-labeled cells) fluorescence of positive cells is equal to that of unlabeled cells. The same has to be done for cells labeled with PE. It is, however, important to note that due to the high possibility of false negativity or false positivity, the use of proper controls (cells stained with only one fluorochrome) for all reagent-cell combinations is often necessary. This is especially true when pairs of fluorochromes with extreme differences in intensity are used.

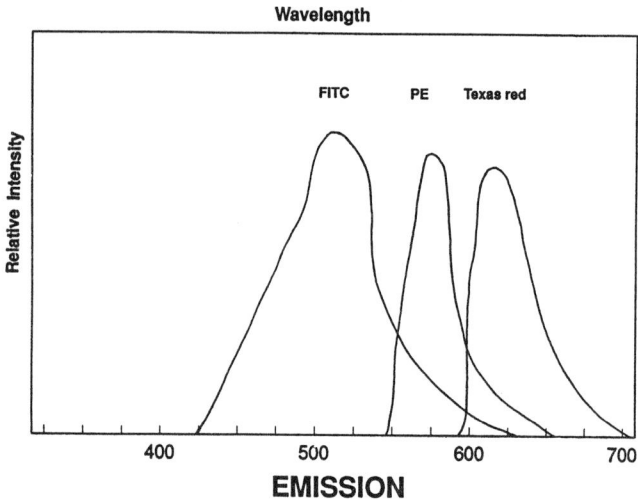

Figure 1. Emission spectra of three stains used in three color immunofluorescence. FITC - fluorescein isothiocyanate; PE - phycoerythrin.

Detector voltage is the voltage used to operate a photomultiplier tube. The level of detector voltage should always be such as to keep both negative and positive cell populations on a histogram.

Forward light scatter (FALS) is measured in front of the path of excitation light, and is a detection of the amount of excitation light scattered by a cell in that particular light beam. Both cell refractive index and cell size influence forward light scatter. FALS is proportional to cell size at narrow forward angle.

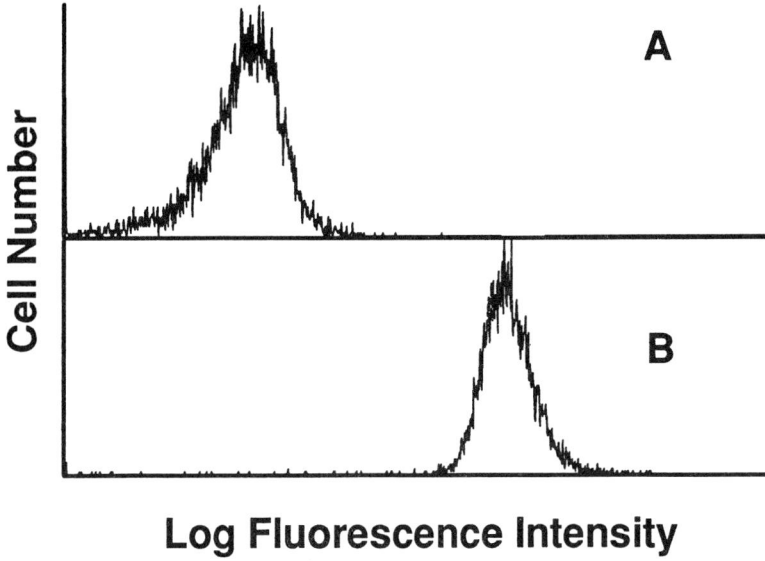

Figure 2. Fluorescence histogram of a negative (A) and positive (B) cell population.

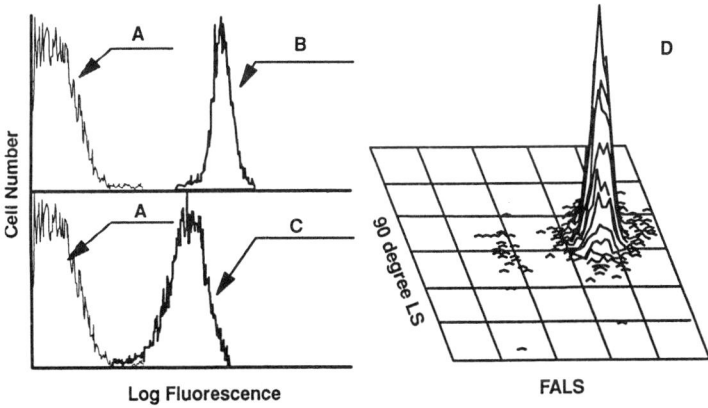

Figure 3. Fluorescence histograms. A - negative cells; B - positive cells; C - the same positive cells evaluated using a lower fluorescence intensity. D - cellular fluorescence and light scatter histogram of the cell population shown as B.

31

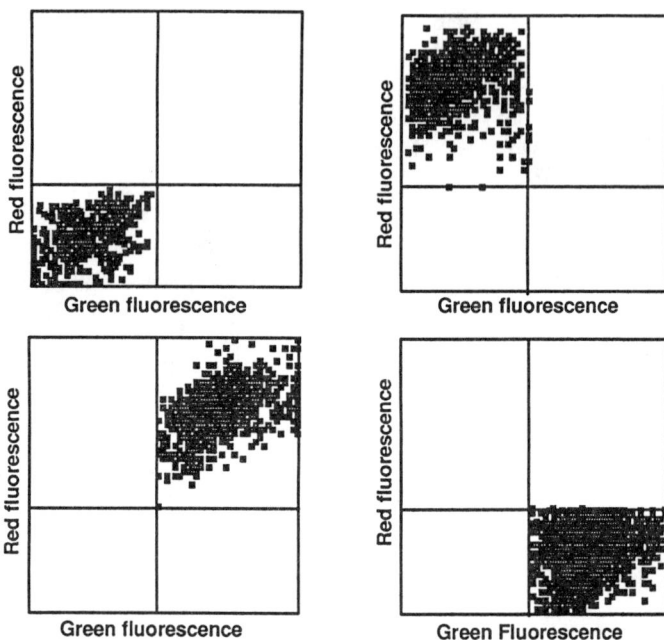

Figure 4. Dual-parameter contour plots of cells labeled with two different antibodies conjugated to FITC and PE, resp. Clockwise, starting in upper left corner: FITC-positive population; FITC- and PE-positive population; FITC- and PE-negative population; PE-positive population. FITC-fluorescein isothiocyanate; PE-phycoerythrin.

It is more common to use a dual-angle light scatter combination, usually forward-angle light scatter and 90° light scatter.

Gates define the parameters for the cells of interest. On a two-parameter contour histogram, you can draw a box defining the region (cell type) of interest.

Regardless of the detection system, the correct use of optical **filters** is critical. Two types of filters are used: absorbance filters and interference (dichroic) filters. Absorbance filters absorb short wavelengths and pass only long wavelengths. Interference filters, on the other hand, are coated such that they are selective mirrors that block by reflecting certain wavelengths and passing other wavelengths. In practice, both types of filters are used simultaneously to block all unwanted light.

Bit map is a part of the contour diagram, which has been selected for further study. Thus, the desired (i.e., fluorescence) parameter(s) will be measured for each bit map.

List mode is a form of data storage. This type of storage allows retrospective analysis of stored data, but requires huge storage capacity. **Calibration** is the safest way to achieve consistency in day-to-day operation of a flow cytometer. It is important to check the machine by a quick analysis of a sample of fluorescent microspheres for which the coefficient of variation was determined just after the machine was optimally calibrated by a serviceman.

There are many other applications of cell cytometers, such as cell sorting, mitochondrial stains, measurement of intracellular pH, glutathione, membrane fluidity, membrane potential, phagocytosis, virus adsorption, pinocytosis, cell killing or enzyme analysis. Most of these methods reach far above immunology and readers seeking more detailed description of both principles of flow cytometry and all possible applications should read some of the comprehensive reviews.[1-5]

A. LABELING OF ANTIBODIES WITH FLUORESCEIN ISOTHIOCYANATE (FITC)

Materials and Reagents

Ascites fluid or serum
$(NH_4)_2SO_4$ or Na_2SO_4
Dimethylsulfoxide (DMSO; Sigma)
5 mg/ml FITC in DMSO (prepare fresh before use)
FITC labeling buffer
Dialysis buffer
5-ml glass tube
Dialyzing tube

Protocol:

1. Purify IgG from monoclonal ascites or serum by $(NH_4)_2SO_4$- or Na_2SO_4-precipitation (see Chapter 10) and dialyze overnight at 4°C against FITC labeling buffer, pH 9.2. Check the concentration (A_{280} x 0.74 x dilution = mg/ml) and adjust it to 1 to 2 mg/ml.
2. Add 20 µl of diluted FITC (in a glass tube) per 1 mg of antibody and incubate 2 hours at room temperature.
3. Dialyze the unbound FITC by dialysis against dialysis buffer. Check the concentration and determine FITC/protein ratio.
4. Measure the optical density at A_{280} and A_{492}.

Protein concentration in mg/ml = A_{280} - (A_{492} x 0.35) / 1.4

Protein in moles = protein mg/ml / 1.5 x 10^5

FITC in moles = A_{492} / 6.69 x 10^5

FITC/protein ratio = moles of FITC / moles of protein

Comments:

1. As an alternative to the dialysis, you can remove the unbound FITC by passage through a Sephadex G-25 column.
2. 1.5 x 10^5 and 0.69 x 10^5 values in the calculation of FITC/protein ratio are the molecular weight of immunoglobulin and the molar extinction coefficient of FITC, respectively.
3. The optimal FITC/protein ratio for flow cytometry is usually 2 to 5:1.

B. LABELING OF ANTIBODIES WITH BIOTIN

Materials and Reagents

Ascites fluid or serum
$(NH_4)_2SO_4$ or Na_2SO_4
Dimethylsulfoxide (DMSO) or dimethyl formamide (DMF; Fisher)
NaN_3
5-ml glass tube
Bovine serum albumin (BSA)
0.1 M $NaHCO_3$, pH 8.5
D-Biotin-N-hydroxysuccinimide ester (BNHS; Boehringer Mannheim #732-494)
Dialyzing tube

Protocol:

1. Purify IgG from monoclonal ascites or serum by $(NH_4)_2SO_4$- or Na_2SO_4-precipitation (see Chapter 10) and dialyze against 0.1 M $NaHCO_3$, pH 8.5. Adjust concentration to 1 mg/ml.
2. Dissolve BNHS in DMSO at 1 mg/ml in a glass tube.
3. Add 200 µl of DMSO solution to 800 µl of antibody (250 µg/ mg protein) and incubate for 1 hour at room temperature.
4. Dialyze against PBS containing 0.05% of NaN_3 and 0.1% BSA.

Table 1

Common Dyes Used in Flow Cytometry

Dye	Ex$_{min}$	Ex$_{max}$	Laser Line	Emission range
DAPI	365	450	351-364	380-420
AO	500	530 (DNA)	488	510-530
AO	500	640 (RNA)	488	590-640
Ethidium	490	610	514, 488	550-640
PI	490	625	514, 488	550-640
FITC	494	517	488	510-530
PE	460	560	488	550-600
RITC	545	600	514	600-620
Lucifer yellow	280-430	540	351-364	500-550
Texas red	595	615	610	620-630
TRITC	460	555	488	530-570
INDO-1	352	450	351-364	400-580
Fura-2	362	340	351-364	300-400
Fluo-3	500	610	488	520-540

C. ALTERNATE METHOD FOR BIOTINYLATION OF ANTIBODIES

Materials and Reagents

D-Biotin-N-hydroxysuccinimide ester (BNHS; Boehringer-Mannheim #732-494)
Dimethyl formamide (DMF)

35

Glass tube
Ascites fluid or serum
$(NH_4)_2SO_4$ or Na_2SO_4
Borate-buffered saline, pH 8.4
Dialyzing tube

Protocol:

1. Purify IgG from mouse monoclonal ascites by $(NH_4)_2SO_4$- or Na_2SO_4- precipitation (see Chapter 10) and dialyze against BBS pH 8.43. Adjust concentration to 3 mg/ml.
2. Dissolve BNHS ester in dimethyl formamide (1.1 mg BNHS/100 µl DMF) in a **glass** tube.
3. Add 25 µl BNHS/DMF to 0.5 ml antibody (183 µg/ mg protein).
4. Incubate 7 hours at room temperature.
5. Dialyze against BBS and render 0.05% in NaN_3.

Comments:

1. The entire procedure can be performed in the dialysis tubing. Clamp tubing at first dialysis. Add BNHS directly to dialysis bag, re-clamp, and incubate in a moist environment (Petri dish with wet gauze) at room temperature. Redialyze in the same tubing.

D. STAINING OF CELLS FOR DETECTION OF SURFACE ANTIGENS

Materials and Reagents

Cell suspension
Labeled or unlabeled antibody
Labeled secondary antibody (in case of unlabeled primary antibody)
100 µg/ml propidium iodide
5-ml centrifuge tubes
RPMI 1640 medium with 5% FCS
Flow cytometer
Ice

Protocol:

1. Put 0.1 ml of cells (1 x 10^5) into centrifuge tubes.

2. Add 10 μl of appropriately diluted antibody to each tube and incubate for 30 minutes on ice.
3. Wash the cells twice by centrifugation at 300 x g for 10 minutes on ice. Discard supernatant and resuspend cell pellet in 0.1 ml of cold RPMI 1640 medium with 5% FCS.
4. In case of unlabeled primary antibodies, repeat steps 2 and 3 with labeled secondary antibodies.
5. Run the samples on the flow cytometer.

Comments:

1. Add 10 μl of diluted propidium iodide simultaneously with last antibody.
2. Do not use propidium iodide with antibodies labeled with phycoerythrin.
3. Always use a negative control. The adequate negative control is either irrelevant antibody of the same isotype as antibody used (for direct labeling) or secondary labeled antibody only.
4. Centrifuge all antibodies at 10,000 x g for 20 minutes prior to use in order to remove all aggregates.

E. STAINING OF CELLS FOR DETECTION OF SURFACE ANTIGENS IN PLATES[6]

<u>Materials and Reagents</u>

Cell suspension
Labeled or unlabeled antibody
Labeled secondary antibody (in case of unlabeled primary antibody)
100 μg/ml propidium iodide
96-well round-bottom microtiter plates (Costar)
RPMI 1640 medium with 5% FCS
Flow cytometer
BSA
PBS
NaN$_3$
Ice

Protocol:

1. Prepare PBS supplemented with 12% BSA and 10 mM NaN$_3$ (12% BSA/PBS/azide).
2. Put 0.1 ml of cells (1 x 10^5 cells/well) into wells of a 96-well plate.

3. Add 10 µl of appropriatelly diluted antibody to each well, mix and incubate for 30 minutes on ice.
4. Wash the cells once by centrifugation through a 0.1 ml cushion of 12% BSA/PBS/azide.
5. In case of unlabeled primary antibodies, repeat steps 2 and 3 with labeled secondary antibodies.
6. Wash the cells once by centrifugation through a 0.1 ml cushion of 12% BSA/PBS/azide.
7. Run the samples on the flow cytometer.

Comments:

1. Add 10 µl of diluted propidium iodide simultaneously with last antibody.
2. Do not use propidium iodide with antibodies labeled with phycoerythrin.
3. Always use a negative control. The adequate negative control is either irrelevant antibody of the same isotype as antibody used (for direct labeling) or secondary labeled antibody only.
4. Centrifuge all antibodies at 10,000 x g for 20 minutes prior to use in order to remove all complexes.
5. Use this modification only when a large number of samples is to be tested.

F. INFLUX OF Ca^{2+} IONS USING Fluo-3

The role of Ca^{2+} as an intracellular regulator has attracted a great deal of interest in recent years. Under normal conditions, intracellular free Ca^{2+} concentrations are maintained by ATPase-Ca^{2+} pumps localized in the plasma membrane that keep molarity concentrations far below that of the external media. Healthy cells maintain an intracellular level of free Ca^{2+} about 10,000 lower than the extracellular environment.

Measurements of the intracellular Ca^{2+} have been important for elucidation of the central role of Ca^{2+}_i as a trigger of the cellular response to activation.

Materials and Reagents

Flow cytometer
Cells
RPMI 1640 medium
Dimethylsulfoxide (DMSO)

38

Fluo-3 (Molecular Probes)
Water bath with cover
CaCl$_2$
PBS
5-ml plastic disposable tubes with cap

Protocol:

1. Wash cells three times in PBS by centrifugation at 250 x g for 5 minutes at room temperature.
2. Resuspend cells in RPMI 1640 medium to a final concentration of 5 x 10^6 cells/ml.
3. Dissolve one tube of the Fluo-3 in 50 µl of DMSO.
4. Add 10 µl of the dissolved dye to 140 µl of RPMI 1640 medium and mix.
5. Add 90 µl of diluted Fluo-3 dye to each milliliter of cell suspension and mix gently.
6. Incubate the mixture of cells and stain for 45 minutes at 37°C in the dark.
7. Remove 100 µl of suspension and add 400 µl of RPMI 1640 medium supplemented with 1 mM CaCl$_2$. Incubate for 2 minutes at 37°C in the dark before adding the test substance.
8. Run the samples on the flow cytometer.

Comments:

1. This method can be done using a single-laser (i.e., argon 488 nm) flow cytometer without any special filters.
2. Do not forget to measure cell suspension before adding the test substance to obtain background levels of Ca^{2+}.

G. INFLUX OF Ca^{2+} IONS USING Indo-1

Materials and Reagents

Flow cytometer
Cells
Hanks' balanced salt solution with calcium and magnesium
Dimethylsulfoxide (DMSO)
2 mg/ml Indo-1 pentaacetoxymethyl ester (Indo-1; dissolved in DMSO; Molecular Probes)

39

1 mg/ml ionomycin (dissolved in DMSO; Calbiochem)
Water bath
PBS
Calcium-EGTA buffers
5-ml plastic disposable tubes with cap

Protocol:

1. Prepare cell suspension and resuspend cells in Hanks' balanced salt solution with calcium and magnesium to a density of 10^6/ml.
2. Put cells into the tubes and add 2 mg/ml Indo-1 solution to a final concentration of 2 µg/ml. Incubate 30 minutes at 37°C.
3. Centrifuge cells at 200 x g for 10 minutes at room temperature. Gently resuspend the cells in Hanks' balanced salt solution with calcium and magnesium and incubate for 15 minutes at room temperature in the dark.
4. Set up the flow cytometer. Use 395-nm and 500-nm bandpass filters.
5. Resuspend cells in a series of calcium-EGTA buffers and ionomycin.[7]
6. Warm an aliquot of Indo-1 loaded cells for 10 minutes at 37°C. Use 1 x 10^5 cells per time interval.
7. Analyze cells in flow cytometer. Keep the cells at all times at 37°C.

H. THE RESPIRATORY BURST OF NEUTROPHILS[8]

The respiratory burst is a burst of oxidative metabolic activity associated with the recognition and ingestion of opsonized particles by neutrophils.[9,10] This burst can be also induced by various mediators such as phorbol myristate acetate or N-formyl-methionyl-leucyl-phenylalanine, and is considered to be a sign of functional activation of neutrophils. It is initiated through the stimulation of NADPH oxidase, which catalyzes the transfer of electrons from NADPH to oxygen.[10,11] The flow cytometric measurement of respiratory burst activity is done using the nonfluorescent probe 2´, 7´-dichlorofluorescein (DCFH)[12] or diacetyl 2´, 7´-dichlorofluorescein (DCFH-DA)[13] which during the respiratory burst are oxidized to the intensely fluorescent 2´, 7´-dichlorofluorescein.

Materials and Reagents

Freshly isolated neutrophils (see Chapter 1)
Ethanol
Diacetyl 2´, 7´-dichlorofluorescein (Eastman-Kodak)
Phorbol myristate acetate (PMA; Sigma)

40

DMSO
10-ml polystyrene tubes
Water bath with cover
Flow cytometer
Hanks' balanced salt solution

Protocol:

1. Isolate neutrophils and dilute them in Hanks' balanced salt solution to a concentration of 10^7 cells/ml.
2. Incubate the neutrophils in 5 µM DCFH-DA (prepared from 500 µM stock in ethanol) in 10-ml polystyrene tubes for 15 minutes at 37°C in dark.
3. Wash the cells once in Hanks' balanced salt solution by centrifugation at 250 x g for 5 minutes at room temperature and add 100 µl of appropriate dilution of PMA (from stock solution of 1 mg/ml in DMSO). The optimal concentration of PMA is 50 to 100 ng/ml. Incubate for 15 minutes at 37°C in the dark.
4. Place the tube with cells on ice and without washing examine the fluorescence using flow cytometer.

Comments:

1. Stock solution of DCFH-DA can be stored at -20°C for 6 months, the stock solution of PMA at -80°C for 6 months.
2. Use a flow cytometer equipped with a laser providing excitation at 488 nm.
3. PMA is used as a positive control.

I. H₂O₂ PRODUCTION MEASUREMENT USING DIHYDRORHODAMINE 123[14]

Material and Reagents

Freshly isolated neutrophils (see Chapter 1)
Dihydrorhodamine 123 (Molecular Probes)
Propidium iodide
10-ml centrifuge tubes
Flow cytometer
PMA
Hanks' balanced salt solution with 10 mM HEPES

41

Dimethyl formamide (DMF)
PBS

Protocol:

1   Incubate 20 µl of cells (2 x 10⁷ cells/ml) with 10 µl dihydrorhodamine
    123 solution (1:10 dilution of stock in Hanks' balanced salt solution
    with 10 mM HEPES; stock is 346 µg/ml of dimethyl formamide) and
    1 ml Hanks balanced salt solution with 10 mM HEPES.
2. Incubate for 5 minutes at 37°C.
3. Add 10 µl of PMA (to a final concentration 100 nM) to a positive
    control tube. Continue incubation. Add 5 µl of propidium iodide (1
    mg/ml in PBS) to counterstain dead cells.
4. Take 250-µl aliquots at 5, 10, and 15 minutes of incubation.
5. Use flow cytometer equipped with an argon laser providing excitation
    at 488 nm.

Comments:

1. The intracellular oxidization of the nonfluorescent dihydrorhodamine
    123 to a green fluorescent rhodamine 123 is a hydrogen peroxide-
    dependent reaction.

J. LACTOFERRIN ASSAY[8,15]

This assay measures release of lactoferrin from granules as a function
of neutrophil activation.

Materials and Reagents

Freshly isolated neutrophils (see Chapter 1)
Hanks' balanced salt solution with Ca^{2+}, Mg^{2+}, and 1% BSA (HBSS/1%
    BSA)
PBS with 0.1% BSA
4% paraformaldehyde in PBS
Anti-human lactoferrin F(ab) conjugated with FITC (Nordic Immunological)
PMA
Flow cytometer
Eppendorf centrifuge tubes
15-ml conical centrifuge tubes

42

Protocol:

1. Resuspend freshly isolated neutrophils at 1×10^6 cells/ml in HBSS/1% BSA and put the suspension on ice.
2. Warm cells at 37°C for 10 minutes.
3. Put 50 μl of PMA (100 nM) as a positive control and 50 μl of tested substance into the appropriate Eppendorf centrifuge tubes and warm for 5 minutes at 37°C.
4. Add 450 μl of cells to each tube, mix and incubate at 37°C for 10 minutes.
5. Add 500 μl of 4% paraformaldehyde, mix and incubate for 20 minutes at room temperature. Wash once by centrifugation at 500 x g for 5 minutes at 4°C with ice-cold PBS with 0.1% BSA.
6. Resuspend cells in 45 μl ice-cold PBS with 0.1% BSA.
7. Add 5 μl of diluted anti-lactoferrin antibody, mix and incubate for 30 minutes at 4°C.
8. Wash twice by centrifugation at 500 x g for 5 minutes at 4°C with ice-cold PBS with 0.1% BSA. Resuspend in 500 μl ice-cold PBS with 0.1% BSA.
9. Add 150 μl of 4% paraformaldehyde.
10. Run on flow cytometer equipped with an argon laser providing excitation at 488 nm.

K. DNA CELL CYCLE ANALYSIS USING PROPIDIUM IODIDE[2,16]

Materials and Reagents

Propidium iodide solution
Cell suspension
5-ml centrifuge tubes
RNAse (Sigma)
Nonidet P-40 (Sigma)
PBS
Flow cytometer
Ice

Protocol:

1. Resuspend 1×10^6 cells/ml in PBS with 50 μg/ml propidium iodide solution. Remove 1 ml of this solution to another tube. Add 1 ml of propidium iodide solution to this second tube.

43

2. Incubate for 30 minutes at 4°C
3. Read on flow cytometer at 488 nm.

L. ALTERNATE METHOD FOR DNA CELL CYCLE ANALYSIS USING PROPIDIUM IODIDE[16]

Materials and Reagents

Propidium iodide solution
Cell suspension
5-ml centrifuge tubes
Flow cytometer
0.1% Na citrate

Protocol:

1. Mix 1×10^6 cells with 1 ml of propidium iodide solution (5 µg/ml of 0.1% Na citrate) and incubate for 5 minutes at 4°C.
2. Read on flow cytometer at 488 nm.

M. DNA CELL CYCLE ANALYSIS USING MITHRAMYCIN[17]

Materials and Reagents

Flow cytometer
Mithramycin (Sigma)
Ethanol
$MgCl_2 \cdot 6H_2O$
5-ml centrifuge tubes
Cell suspension

Protocol:

1. Resuspend 1×10^6 cells in 1 ml of 25% ethanol containing 100 µg/ml mithramycin and 15 mM $MgCl_2$.
2. Incubate on ice for 20 minutes.
3. Read on flow cytometer equipped with an argon laser at 488 nm. Use the filter for phycoerythrin.

REFERENCES

1. **Wood, J. C. S., Horton, A. F., Byrne, J. D., Pedroso, R. I., Bisnow, M. and Auer, R.** E., Dual laser excitation flow cytometry: the state-of-the-art, in *Flow Cytometry: Advanced Research and Clinical Applications, Vol.1,* Yen, A., Ed., CRC Press, Boca Raton, 1989, 5.

2. **Robinson, J.P.,** Ed., *Handbook of Flow Cytometry Methods,* Wiley-Liss, New York, 1993.

3. **Melamed, M. R., Lindomo, T. and Mendelson, M. L.,** Eds., *Flow Cytometry and Sorting,* 2nd ed., Willey-Liss, New York, 1990.

4. **June, C. H. and Rabinovitch, P. S.** Immunofluorescence and cell sorting, in *Current Protocols in Immunology,* Coligan, J. E., Kruisbeek, A. M., Margulies, D. H., Shevach, E. M. and Strober, W. Eds., Green Publishing and Wiley-Interscience, New York, 1991, 5.0.1.

5. **McLean Grogan, W. and Collins, J. M.,** *Guide to Flow Cytometry Methods,* Marcel Dekker, New York, 1990.

6. **Muto, S., Vetvicka, V. and Ross, G.D.,** CR3 (CD11b/CD18) expressed by cytotoxic T cells and NK cells is upregulated in a manner similar to neutrophils following stimulation with various activating agents, *J. Clin. Immunol.* 13, 175, 1993.

7. **Chused, T. M., Wilson, H. A., Greenblatt, D., Ishida, Y., Edison, L. J., Tsien, R. Y. and Finkelman, F.D.,** Flow cytometric analysis of murine splenic B lymphocyte cytosolic free calcium response to anti-IgM and anti-IgD, *Cytometry,* 8, 396, 1987.

8. **Kenny, P. A. and Finlay-Jones, J. J.,** Flow-cytometric measurement of respiratory burst activity of neutrophils, in *Laboratory Methods in Immunology,* Zola, H., Ed., CRC Press, Boca Raton, 1990, 131.

9. **Babior, B. M.,** Oxygen-dependent microbial killing by phagocytes, *N. Engl. J. Med.,* 298, 659, 1978.

10. **Rossi, F., Dri, P., Bellavite, P., Zabucchi, G. and Berton, G.,** Oxidative metabolism of inflammatory cells, *Adv. Inflammation Res.,* 1, 139, 1979.

11. **Forman, H. J. and Thomas, M. J.,** Oxidant production and bactericidal activity of phagocytes, *Annu. Rev. Physiol.,* 48, 669, 1986.

12. **Bass, D. A., Parce, J. W., DeChatelet, L. R., Szejda, P., Seeds, P., Seeds, M. C. and Thomas, M.,** Flow cytometric studies of oxidative product formation by neutrophils: a grade response to membrane stimulation, *J. Immunol.,* 130, 1910, 1983.

13. **Brandt, R. and Keston, A. S.,** Synthesis of diacetyldichlorofluorescein: a stable reagent for fluorimetric analysis, *Anal. Biochem.,* 11, 6, 1965.

14. **Rothe, G., Emmendorffer, A., Oser, A., Roesler, J. and Valet, G.,** Flow cytometric measurement of the respiratory burst activity of phagocytes using dihydrorhodamine 123, *J. Immunol. Methods,* 128, 133, 1991.

15. **Butler, T. W., Heck, L. W., Huster, W. J., Grossi, C. E. and Barton, J.C.,** Assesment of total immunoreactive lactoferrin in hematopoietic cells using flow cytometry, *J. Immunol. Methods,* 108, 159, 1988.

16. **Krishan, A.,** Rapid flow cytometric analysis of mammalian cell cycle by propidium iodide staining, *J. Cell. Biol.,* 66, 188, 1975.

17. **Tobey, R. A. and Crissman, H. A.,** Unique techniques for cell cycle analysis utilizing mitramycin and flow microfluorometry, *Exp. Cell Res.,* 93, 235, 1975.

Chapter 3

PROLIFERATION ASSAYS

The assessment of cellular proliferation is perhaps one of the most often used techniques in cellular immunology. Mitogenic activation or specific recognition of antigen by lymphocytes results in the elicitation of a sequence of events that culminate with entry of the cells into cycle and cell division. In addition, pre-activated lymphocytes or a variety of cell lines are able to respond to growth-promoting cytokines with increased proliferation. Because of its simplicity, the measurement of cellular proliferation is routinely used as a convenient way to determine:

a)   the magnitude of lymphocytic responses to antigens and/or mitogens;
b)   the production and concentration of growth-promoting cytokines (bioassays);
c)   cellular toxicity.

Although cellular proliferation can be assessed by determining the increase in viable cell numbers through the actual enumeration of cells by direct microscopy or automated cell counting, this approach can be time-consuming, especially when working with a large number of samples or with mixed cell cultures. Instead, the two most commonly used techniques for the purpose of measuring cellular proliferation are

1)   incorporation of titriated [^{3}H] thymidine into newly synthesized DNA; and
2)   reduction of tetrazolium dyes, such as 3-[4,5-dimethylthiazol-2-yl]-2,5-diphenyl tetrazolium bromide (MTT), by active mitochondria.

These two assays are described below.

I. INCORPORATION OF TRITIATED THYMIDINE INTO DNA

As cells enter the "S" phase of the cell cycle, chromosome replication takes place, with the incorporation of soluble nucleotide precursors into newly synthesized DNA. In this assay, dividing cells are incubated ("pulsed") with radioactive [^{3}H] thymidine for several hours, after which the amount of

47

radioactivity incorporated into their DNA is determined by harvesting the cells onto glass fiber filters followed by liquid scintillation counting.

Materials and Reagents

Cells (primary culture, indicator cell lines, etc.)
Culture medium (depending on the cell culture; for many applications the culture medium of choice is RPMI 1640 medium supplemented with 10% FCS, 100 U/ml penicillin, 100 µg/ml streptomycin, 2 mM L-glutamine and 50 µM 2-mercaptoethanol)
[^{3}H-methyl] thymidine (approx. 20 to 50 Ci/mmol), sterile (e.g., Amersham, Du Pont NEN)
96-well tissue culture plates (flat bottom)
CO_2 incubator (humidified, set at 37°C and 5% CO_2)
Multichannel pipettete or repeating dispenser
Cell harvester (Skatron Instruments)
Glass fiber filter mats (Skatron Instruments)
Liquid scintillation vials (Research Products International)
Liquid scintillation cocktail (e.g., Econo-Safe™, Research Products International)
Liquid scintillation counter

Protocol:

1. Plate the indicator cells in 96-well plates in a final volume of 0.1 to 0.2 ml. Normally, each culture condition (e.g., dilution) should be tested in triplicate. Include a background proliferation control in which the cells are cultured in medium alone. Incubate at 37°C in a CO_2 incubator for the appropriate length of time.

2. Dilute the [^{3}H-methyl] thymidine stock solution with culture medium to a concentration of 50 µCi/ml. With the aid of a multichannel pipettete, add 20 µl of the diluted solution to each well. This gives a final concentration of 1 µCi per well.

3. Return the plates to the incubator. Incubate for additional 4 to 18 hours. (With rapidly dividing cells, a 4- to 6-hour "pulse" gives adequate results.)

4. Harvest the cells onto glass fiber mats using a cell harvester and distilled water.

5. Allow the filters to air-dry. Place individual filters into liquid scintillation vials and add 2 to 3 ml of liquid scintillation cocktail.

6. Count in liquid scintillation counter. Express proliferation data based on cpm or dpm per culture. Alternatively, proliferation data can

48

be expressed as the "Stimulation Index", calculated by dividing the proliferation of the stimulated cells (in cpm) by the background proliferation (in cpm).

Comments:

1. The culture medium, the final cell density and the length of incubation will depend on the nature of the indicator cells and on the nature of the stimulus. Generally, for indicator cell lines (i.e., HT-2, CTLL) which divide rapidly, cell densities of 0.5 to 1 x 10^5/ml give adequate results. Freshly isolated lymphocytes or mononuclear cells, on the other hand, require higher cell densities (0.5 to 2 x 10^6/ml). As for incubation times, measurement of proliferation of indicator cell lines to growth-promoting cytokines requires incubations of no more than 48 hours, whereas proliferation of lymphocytes to antigens or mitogens may require incubations as long as 72 to 96 hours.

2. It is important to use tritiated thymidine of high purity and correct specific activity. It is recommended to use thymidine labeled in the 5-methyl position in order to avoid the potential risk of labeling RNA due to conversion of thymidine into uridine by demethylation. [^{3}H-methyl] thymidine is commercially available at various specific activities, ranging from 2 to 100 Ci/mmol. The preparations with the lower specific activities are available sterile, in aqueous solution. Ethanol, however, is usually added to the preparations of higher specific activity as a free radical scavenger in order to reduce radiation decomposition. When using these preparations, the investigator should be aware of the final ethanol concentration, as it may be toxic to some cells.

3. Many liquid scintillation formulations are adequate for this purpose; however, because of the relatively large quantities of radioactive waste generated, the use of a biodegradable liquid scintillation cocktail is recommended.

4. Disintegrations per minute (dpm) = counts per minute (cpm) ÷ counting efficiency

II. NONRADIOACTIVE DETECTION METHODS (MTT ASSAY)

Due to the potential hazards associated with the use of radioactive substances, many investigators prefer the use of nonradioactive detection methods for the measurement of cellular proliferation. The most common of these methods is known as the "MTT Assay".[1] Such technique is based on the cleavage of a yellow tetrazolium dye (3-[4,5-dimethylthiazol-2-yl]-2, 5-

diphenyl tetrazolium bromide, MTT) into insoluble purple formazan by dehydrogenases in active mitochondria. Dead cells are unable to perform this reaction. In this assay, an MTT solution is added to the dividing cells and, after a 4-hour incubation period, the amount of purple formazan generated is spectrophotometrically determined using a multiwell spectrophotometer or "ELISA reader".

Materials and Reagents

Cells (primary culture, indicator cell lines, etc.)

Culture medium (depending on the cell culture; for many applications the culture medium of choice is: RPMI 1640 medium supplemented with 10% FCS, 100 U/ml penicillin, 100 µg/ml streptomycin, 2 mM L-glutamine and 50 µM 2-mercaptoethanol)

96-well tissue culture plates (flat bottom)

CO_2 incubator (humidified, set at 37°C and 5% CO_2)

Multichannel pipettete or repeating dispenser

MTT (Sigma Cat. No. M-5655). Prepare a stock solution of MTT (5 mg/ml) in PBS. Filter.

Acidic isopropanol (0.04 N HCl in isopropanol)

Multiwell spectrophotometer (ELISA reader)

Protocol:

1. Plate the cells in 96-well tissue culture plates as described in the previous section, including cells cultured in medium alone as a background proliferation control (triplicates are recommended). Culture in a CO_2 incubator at 37°C for the appropriate length of time.

2. Using a multichannel pipettete, add 10 µl of the MTT stock solution per 100 µl of culture medium. (Final MTT concentration is 0.5 mg/ml.)

3. Incubate at 37°C for 4 hours.

4. If the cells are nonadherent, add 100 µl/well of the acidic isopropanol solution directly onto the wells. If the cells are adherent, the medium can be discarded first, and then the acidic isopropanol (100 µl per well) can be added directly onto the cells.

5. Mix well and incubate at room temperature until all crystals are dissolved (5 to 10 min).

6. Read on a multiwell spectrophotometer using a test wavelength of 570 nm and a reference wavelength of 630 nm. Read plates within one hour of addition of the acidic isopropanol.

Comments:

1. Red blood cells do not cleave MTT to a significant extent nor do they interfere with the assay up to concentrations of 2 x 10^6 cells/ml.[1]

2. Several modifications to the original assay described by Mosmann[1] have been reported. Most of these modifications address the problem of protein precipitation, especially when using culture medium containing high serum concentrations, during the formazan-extraction step using acidic isopropanol.[2] Sodium dodecyl sulfate (SDS) appears to minimize protein precipitation but requires longer extraction times, whereas isopropanol accelerates the extraction but leads to increased development of turbidity due to protein precipitation. Extraction reagents in some of these modifications include:[2]

   a) 10% SDS buffered to pH 4.7 with acetate buffer
   b) 10% SDS-50% isopropanol - 0.01 N HCl, pH 5.5
   c) 20% SDS in 50% N,N-dimethylformamide buffered to pH 4.7 with acetic acid
   d) 3% SDS in acidified (0.04 N HCl) isopropanol

   In another modification resulting in improved sensitivity and reproducibility,[3] the normal medium is removed before the addition of MTT. The incubation with MTT is then carried out in serum-free medium devoid of phenol red, thus avoiding potential precipitation of proteins and the need for acidification. The purple formazan product is then dissolved using propanol or ethanol. This assay uses 560 and 690 nm as the test and reference wavelengths, respectively.

3. Another nonradioactive detection assay for cell proliferation is available from Amersham. This assay is based on the incorporation of 5-bromo-2' deoxyuridine (BrdU), a thymidine analog, into DNA, followed by detection of incorporated BrdU using a specific peroxidase-labeled monoclonal antibody against BrdU.

REFERENCES

1. **Mosmann, T.,** Rapid colorimetric assay for cellular growth and survival: application to proliferation and cytotoxicity assays, *J. Immunol. Methods,* 65, 55, 1983.

2. **Niks, M. and Otto, M.,** Towards an optimized MTT assay, *J. Immunol. Methods,* 130, 149, 1990.

3. **Denizot, F. and Lang, R.,** Rapid colorimetric assay for cell growth and survival. Modifications to the tetrazolium dye procedure giving improved sensitivity and reliability, *J. Immunol. Methods,* 89, 271, 1986.

Chapter 4

PREPARATION OF CELL CLONES

I. PREPARATION OF MOUSE T CELL CLONES

A. Th CLONES REACTIVE WITH SOLUBLE ANTIGENS[1]

Antigen-specific T cells can usually be cloned into two distinct populations of Th cells: Th1 cells, which produce IL-2 and IFNγ, and Th2 cells, which produce IL-4 and whose proliferation is inhibited by IFNγ.[2] The addition of either ConA-supernatant (containing high levels of IL-2) or MLC-supernatant (containing IL-2, IL-4, and IFNγ) during the cloning process results in the preferential isolation of Th2 or Th1 clones, respectively.

Materials and Reagents

Mice
Antigen
Freund's complete adjuvant (Sigma)
PBS
DMEM medium with 5% FCS, antibiotics
Irradiated syngeneic spleen cells
ConA-supernatant or MLC-supernatant (see Sections B.1 and B.2)
96-well flat-bottom tissue culture plates
24-well flat-bottom tissue culture plates
Human rIL-2
Mouse rIFNγ
Humidified 37°C, 5% CO_2 incubator

Protocol:

1. Inject mice subcutaneously with a 1:1 emulsion of antigen (50 µg/ml to 1000 µg/ml) in PBS and Freund's complete adjuvant into both hind foot pads and the base of the tail.
2. Seven days later remove popliteal, periaortic, and inguinal lymph nodes and spleen (see Chapter 1) and prepare single-cell suspension of lymph node cells and spleen cells. Wash the cells twice by centrifugation at 350 x g at 4°C in complete medium and count them.

3. Irradiate (2,000 to 4,000 rad) spleen cells.
4. Culture 2×10^6 lymph node cells for 8 days with 6×10^6 irradiated syngeneic spleen cells with antigen (50 µg/ml to 500 µg/ml) in 1.5 ml of complete DMEM medium in each well of a 24-well plate in a humidified 37°C, 5% CO_2 incubator.
5. Add to each well of a 96-well flat-bottom tissue culture plate:
   50 µl irradiated syngeneic spleen cells (2×10^7 cells/ml)
   50 µl antigen (200 µg/ml to 1,600 µg/ml)
   25 µl human rIL-2 (80 U/ml)
   25 µl mouse rIFNγ (4,000 U/ml)
6. Pool lymph node cells (from step # 4), wash once in complete DMEM medium at 4°C and dilute them in complete DMEM medium at 1×10^3 cells/ml. Add 50 µl of these cells into each well of the plate prepared in step 5 and incubate 7 days in a humidified 37°C, 5% CO_2 incubator.
7. Aspirate 50 µl of media and add 25 µl human rIL-2 (80 U/ml) and 25 µl mouse rIFNγ (4,000 U/ml) to each well. Incubate for additional 5 to 7 days in a humidified 37°C, 5% CO_2 incubator.
8. Check under the microscope. Use only the wells with a single cluster of growing T cells.

Comments:

1. You can use 100 µl of 50% MLC-supernatant instead of rIL-2 and rIFNγ.
2. This protocol results in production of Th1 cells. To prepare Th2 cells, add 50 µl of human rIL-2 (40 U/ml) or 50 µl 40% ConA-supernatant (instead of rIFNγ) at steps # 4 and # 6.

B. MAINTAINING Th CLONES

Protocol:

1. Add the following into a 24-well flat-bottom tissue culture plate:
   5×10^4 to 2×10^5 cloned T cells in 100 µl of complete DMEM medium;
   6×10^6 irradiated syngeneic spleen cells in 0.9 ml of complete DMEM medium;
   50 to 500 µg/ml of antigen;
   for Th1 clones: human rIL-2 (25 U/ml) and mouse rIFNγ (250 U/ml) or 33% MLC-supernatant;
   for Th2 clones: human rIL-2 (25 U/ml) or 10% ConA-supernatant.

2. Cultivate cells in a humidified 37°C, 5% CO_2 incubator and subculture them every 7 days.

Comments:

1. The concentrations of growth factors are final concentrations. For long-term cultivation, use 10 U/ml of IL-2.

1. ConA-Supernatant

Materials and Reagents

Mouse spleen cells
Complete DMEM medium with 5% FCS
FCS
50-ml conical centrifuge tubes
25-cm^2 tissue culture flasks
Concanavalin A (ConA; Sigma)
Humidified 37°C, 5% CO_2 incubator

Protocol:

1. Cultivate 1.25 x 10^6 spleen cells/ml for 24 to 48 hours with 2.5 µg/ml ConA in DMEM (without MOPS) containing 5% FCS in a humidified 37°C, 5% CO_2 incubator.
2. Collect culture supernatant and centrifuge at 800 x g for 10 minutes at 4°C.
3. Test ConA-supernatant for IL-2 content (see Chapter 6).

Comments:

1. Rat spleen cells can be used instead of murine cells.

2. MLC-Supernatant

Materials and Reagents

C57Bl/6 and DBA/2 mice
DMEM medium with 5% FCS
FCS
50-ml conical centrifuge tubes

25-cm^2 tissue culture flasks
Humidified 37°C, 5% CO_2 incubator

Protocol:

1. Mix 2.5 x 10^7 C57Bl/6 mouse spleen cells with the same number of irradiated (2,000 to 4,000 rad) DBA/2 mouse spleen cells in 20 ml of complete DMEM medium with 5% FCS and incubate in upright 25-cm^2 plastic tissue culture flasks for 14 days in a humidified 37°C, 5% CO_2 incubator.
2. Collect cells in a 50-ml centrifuge tube and centrifuge at 200 x g at room temperature for 10 minutes. Discard supernatant and wash cells once with complete DMEM medium.
3. Mix 6 x 10^6 cells from step 2 with 2.5 x 10^7 irradiated DBA/2 spleen cells in 20 ml of complete DMEM medium with 5% FCS and incubate in an upright 50-ml plastic tissue culture flasks for 36 hours in a humidified 37°C, 5% CO_2 incubator.
4. Collect culture supernatant and centrifuge at 800 x g for 10 minutes at 4°C.
5. Test MLC-supernatant for IL-2 and IFNγ content (see Chapter 6).

II. PREPARATION OF HUMAN T CELL CLONES

Both protocols for the generation of antigen-specific human T cell clones are based on use of mononuclear cells from an antigen-sensitized donor. The existence of natural and/or polyspecific antibodies (for review see reference[3]) makes it possible to generate a human T cell clone even without immunization, but generally the chances are extremely low. The use of antigen-specific bulk cultures overcomes this problem; however, one has to keep in mind that these cell populations are not clonal.

A. USING STIMULATION/STARVING CYCLES

Materials and Reagents

Peripheral blood mononuclear cells (see Chapter 1)
Antigen
RPMI 1640 medium with antibiotics, 25 mM HEPES
RPMI 1640 medium with antibiotics, 25 mM HEPES, and 10% human AB
    serum

56

Irradiated (3,500 rad) autologous mononuclear cells (from the same donor)
Human rIL-2
Ficoll-Hypaque™
24-well flat-bottom tissue culture plates
96-well flat-bottom tissue culture plates
Humidified 37°C, 5% CO_2 incubator

Protocol:

1. Dilute mononuclear cells in RPMI 1640 medium containing 10% human AB serum to 2 x 10^6 cells/ml and add 2 ml of this suspension into wells of a 24-well plate.
2. Add optimal amount of antigen (5 to 500 µg/ml) and incubate 7 days in a humidified 37°C, 5% CO_2 incubator.
3. Pool all cells, and wash once by centrifugation at 450 x g in RPMI 1640 medium at room temperature. Discard supernatant and centrifuge cells through Ficoll-Hypaque™ gradient (see Chapter 1).
4. Recover cells, wash them twice by centrifugation at 450 x g in RPMI 1640 medium at room temperature and count them. Dilute the cells to 10^6 cells/ml in RPMI 1640 medium containing 10% human AB serum.
5. Add 1 ml of this cell suspension into each well of a 24-well plate. Add 1 ml of 3 x 10^6/ml of irradiated mononuclear cells into each well. Incubate 7 days in a humidified 37°C, 5% CO_2 incubator. *This is a starving cycle.*
6. Repeat steps # 3, 4 a 5 with the only exception of adding the optimal concentration of antigen in step 5. *This is a stimulation cycle.*
7. Repeat starving and stimulation cycles three times each.
8. Centrifuge cells through Ficoll-Hypaque™ gradient, wash twice by centrifugation at 450 x g in RPMI 1640 medium at room temperature and evaluate the antigen specificity by an antigen-specific proliferation assay (see Chapter 3).
9. Prepare clones by limiting dilution in 96-well plates containing:
   10^5 irradiated autologous mononuclear cells (per well);
   30 U/ml IL-2;
   Antigen;
   T cells from step # 8 (plate 0.3, 1, and 3 cells per well).
10. Incubate 7 days in a humidified 37°C, 5% CO_2 incubator. Check under the microscope. Use only those wells with a single cluster of growing T cells.
11. Transfer single clusters of growing cells to 24-well plates with irradiated autologous mononuclear cells and antigen and incubate for 7 days in a humidified 37°C, 5% CO_2 incubator.

12. Check the growing cells for antigen specificity (see Chapter 3).
13. Continue cultivation and expansion of antigen-specific clones by repeating starving and stimulation cycles.

Comments:

1. The optimal antigen concentration should be determined before the experiment either by a test of lymphokine production (see Chapter 6) or T cell proliferation (see Chapter 3).
2. The optimal dose for irradiation of mononuclear cells should be determined in a preliminary experiment.

B. ANTIGEN-NONSPECIFIC T CELL CLONES[4]

Materials and Reagents
Peripheral blood mononuclear cells (see Chapter 1)
RPMI 1640 medium with antibiotics, 25 mM HEPES, and 10% human AB serum
Irradiated (3,500 rad) autologous mononuclear cells (from the same donor)
Irradiated (3,500 rad) allogeneic mononuclear cells (from a different donor)
Human rIL-2
Phytohemagglutinin (PHA)
Ficoll-Hypaque
24-well flat-bottom tissue culture plates
96-well flat-bottom tissue culture plates
Humidified 37°C, 5% CO_2 incubator

Protocol:
1. Isolate T lymphocytes from peripheral blood mononuclear cells (see Chapter 1).
2. Using a 24-well flat-bottom tissue culture plate, mix into each well 10^5 T cells with 5 x 10^4 irradiated autologous cells in a final volume of 2 ml complete RPMI medium containing 5 µg/ml PHA. Incubate for 4 days in a humidified 37°C, 5% CO_2 incubator.
3. Aspirate 1 ml of supernatant and add 1 ml of fresh medium supplemented with 50 U rIL-2. Incubate for 3 days in a humidified 37°C, 5% CO_2 incubator.
4. Centrifuge cells through a Ficoll-Hypaque gradient, wash twice by centrifugation at 450 x g in RPMI 1640 medium at room temperature, and count them.
5. Put 10^5 irradiated allogeneic cells, 5 µg/ml PHA, and 25 U/ml rIL-2

(final volume of 200 µl) into each well of a 96-well flat-bottom tissue culture plate. Plate T cells (from step # 4) into each well (plate 0.3, 1, and 3 cells per well). Incubate for several days in a humidified 37°C, 5% CO_2 incubator.

6. Aspirate 100 µl of supernatant and add 100 µl of complete medium supplemented with 50 U/ml rIL-2. Incubate for an additional 3 to 4 days in a humidified 37°C, 5% CO_2 incubator.

7. Aspirate 100 µl of supernatant and add 5 x 10^4 irradiated allogeneic cells, 5 µg/ml PHA and 50 U/ml rIL-2 (final volume 200 µl). Incubate for additional 4 days in a humidified 37°C, 5% CO_2 incubator.

8. Repeat step # 6. At the end of incubation, identify wells with growing clones under a microscope and expand growing cultures by repeating steps # 6 and # 7.

C. LONG-TERM MAINTENANCE OF ANTIGEN-SPECIFIC AND NONSPECIFIC T CELL CLONES

Materials and Reagents

Ficoll-Hypaque™
T cell clones
RPMI 1640 medium with antibiotics, 25 mM HEPES
RPMI 1640 medium with antibiotics, 25 mM HEPES with 10% human AB
    serum
Human rIL-2
PHA
24-well flat-bottom tissue culture plates
Irradiated (3,500 rad) allogeneic mononuclear cells (from a different donor)
Humidified 37°C, 5% CO_2 incubator

Protocol:

1. Centrifuge cells through a Ficoll-Hypaque™ gradient, wash twice by centrifugation at 450 x g in serum-free RPMI 1640 medium at room temperature, count them, and resuspend at 10^6 cells/ml in complete RPMI 1640 medium.

2. Plate 1 ml of this suspension into each well of a 24-well plate.

3. Add 1 ml irradiated allogeneic mononuclear cells (2 x 10^6 cells/ml) with 5 µg/ml PHA and 15 U/ml rIL-2 into each well. Incubate the plate for 4 days in a humidified 37°C, 5% CO_2 incubator.

4. Aspirate 1 ml of supernatant and add 1 ml of fresh complete RPMI 1640 medium supplemented with 25 U/ml rIL-2. Incubate for an additional 3 days in a humidified 37°C, 5% CO_2 incubator.
5. Pool the growing T cells. Centrifuge pooled cells through Ficoll-Hypaque™ gradient, wash twice by centrifugation at 450 x g in serum-free RPMI 1640 medium at room temperature, count them, and resuspend at 10^6 cells/ml in complete RPMI 1640 medium.
6. Repeat steps # 3, 4 and 5 every week.

Comments:

1. The above-described protocol is for antigen-nonspecific T cell clones. To maintain antigen-specific T cell clones, stimulate them once a month with irradiated autologous mononuclear cells and antigen. Incubate the mixture of T cells and irradiated cells (at a ratio of 1:3) with the antigen for 7 days in a humidified 37°C, 5% CO_2 incubator.

D. BULK CULTURES OF ANTIGEN-SPECIFIC T HELPER CELLS[5]

Materials and Reagents

Peripheral blood mononuclear cells
Human rIL-2
AIM V medium (Gibco)
AIM V medium supplemented with 2.5% autologous serum
Antigen
Ficoll-Hypaque™
75-cm² tissue culture flasks
Humidified 37°C, 5% CO_2 incubator

Protocol:

1. Incubate 2 x 10^7 peripheral blood mononuclear cells in serum-free AIM V medium supplemented with 30 µg/ml of antigen (total volume of 25 ml/flask) for 3 days in a humidified 37°C, 5% CO_2 incubator.
2. Pool nonadherent cells, isolated by washing the culture flasks three times with AIM V medium, and wash them once by centrifugation in AIM V medium for 10 minutes at 450 x g at room temperature.
3. Incubate these cells for an additional 14 days in AIM V medium with 2.5% autologous serum, 30 µg/ml of antigen and 30 U/ml rIL-2 (total volume of 25 ml/flask) in a humidified 37°C, 5% CO_2 incubator.
4. Centrifuge cells through Ficoll-Hypaque™ gradient, wash twice by

60

centrifugation at 450 x g in desired medium at room temperature, count them, and resuspend for further experiments.

III. PREPARATION OF HUMAN B CELL CLONE[6]

Materials and Reagents

Marmoset B95-8 cell line releasing EVB (CRL-1612; ATCC)
RPMI 1640 medium with 10 % FCS
Cyclosporin A (Sandoz)
96-well flat-bottom tissue culture plates (Costar)
25-cm^2 tissue culture flasks (Costar)
Ethanol
Human rIL-2
15-ml conical centrifuge tubes
0.22-µm filter (Costar)
Humidified 37°C, 5% CO_2 incubator

Protocol:

1. Incubate the B95-8 cell line in tissue culture flasks in RPMI 1640 medium containing 10% FCS. Collect the cell-free supernatant and filter it by passing through the 0.22-µm filter.
2. Isolate mononuclear cells from peripheral blood (see Chapter 1). Resuspend 10^7 of washed (at least five times; centrifugation at 300 x g for 5 minutes at 4°C) mononuclear cells in 10 ml RPMI 1640 with 10% FCS, and 30% supernatant of the B95-8 cell line and 600 ng/ml cyclosporin A (dissolve in ethanol at 1 mg/ml).
3. Incubate for 16 hours at 37°C. Wash the cells three times in RPMI 1640 medium with 10% FCS at 250 x g for 10 minutes.
4. Cultivate cells at concentration 5 x 10^4 cells/well in 96-well flat-bottom tissue culture plates in 200 µl RPMI 1640 with 10% FCS and 600 ng/ml cyclosporin A for 13 days.
5. Clone cells by limiting dilution.

Comments:

1. The optimal time for initial incubation with EBV-containing B95-8-derived supernatant may vary.

IV. CLONING BY LIMITING DILUTION TECHNIQUE

In the process of selecting hybridomas, T cell or B cell lines of the desired specificity, a final but very important step is the selection of a clonal cell population. This will insure not only that the antibodies and/or cell lines have a single specificity, but will help prevent the overgrowth of variants that have lost the desired phenotype. In the case of hybridoma selection, for example, positive wells may contain cells derived from more than one clone, or even if derived from a single hybridoma, chromosome loss may result in appearance of a nonsecreting progeny of cells that might overgrow antibody-secreting cells. In addition, subcloning may also allow the selection of high-antibody secreting variants.

Several techniques have been used for cloning purposes. Some of these are

1. **Cloning in soft agar:** In this technique, cells are diluted and cultured in soft-agar plates,[7] so that cell clones will grow as colonies (arising from a single cell precursor) at different locations on the plate. Since the agar prevents the cells from moving and mixing, the clonality of the colonies is insured. This technique, however, present some disadvantages, such as:

a) the cloning efficiency may be relatively low compared to other techniques, such as limiting dilution;
b) the cells grown as colonies need to be adapted to liquid culture;
c) in the case of hybridomas, the cells need to be grown as liquid cultures and supernatants generated before they can be screened.

2. **Limiting dilution:** In this technique, the cells are first diluted and cultured in 96- or 24-well tissue culture plates at cell densities at which, statistically (Poisson distribution), most of the wells would contain only a single cell. Many variations of this basic technique exist, depending on the nature of the cells being clones and their need for stimulation and/or growth factors. For example, several limiting dilution techniques for the cloning of hybridomas differ on their use of "feeder" cells or "conditioned supernatants" from thymocytes, splenic cells or growth factor-producing cell lines (e.g., P388D$_1$ cells and IL-6).[8] On the other hand, cloning of antigen-specific T cells might require the presence of antigen-presenting cells, antigen and IL-2.

3. **Other techniques:** Cloning of positive hybridomas has also been done using the fluorescence-activated cell sorter (FACS).[9] In this technique, the FACS can be used to first select positive hybridomas based on the binding

of a fluorescent antigen, while at the same time, sorting the positive cells onto 96-well tissue culture plates at 1 cell/well.

Because of its simplicity and lack of dependence on sophisticated equipment, only the limiting dilution technique will be further discussed in this chapter. The method described is used for the limiting dilution cloning of hybridomas. Whereas the basic technique is the same, cloning of T cell lines requires different conditions, particularly antigen and accessory cells.

Materials and Reagents:

Candidate cells (primary hybridomas)
Cloning medium: DMEM containing 20% heat inactivated fetal calf serum (FCS), 4.5 g/liter glucose, 100 U/ml penicillin, 100 µg/ml streptomycin, and 20% conditioned medium from SP2/0-Ag14 cells (ATCC Cat. No. CRL 1581)
Culture medium: DMEM containing 10% heat inactivated fetal calf serum (FCS), 4.5 g/liter glucose, 100 U/ml penicillin, 100 µg/ml streptomycin.
96-well tissue culture plates (flat bottom)
15- and 50-ml centrifuge tubes
Multichannel pipettete
Sterile troughs
Pasteur pipettes
Inverted microscope
Centrifuge (set at 4°C)
Incubator (humidified, set at 37°C and 5% CO_2)
Reagents and materials necessary for hybridoma screening

Protocol:

(Note: All reagents and materials must be sterile. Use appropriate sterile technique throughout the procedure.)

1. Resuspend the candidate hybridoma by pipetting gently with the aid of a sterile Pasteur pipette. Transfer the cells to a 15-ml centrifuge tube and assess viability by Trypan blue exclusion.
2. Readjust the cell density to 1×10^5 cells/ml using culture medium. Prepare first a 1:20 dilution of the cells in cloning medium by pipetting 0.1 ml of the original cell suspension and diluting it into 1.9 ml of medium. Next, prepare a 1:100 dilution by pipetting 0.25 ml of this suspension (1:20) and mixing it with 24.75 ml of cloning medium (this suspension should now contain 50 cells/ml). Finally, prepare a 5

cells/ml suspension by diluting 2.0 ml of the 50 cells/ml suspension and mixing it with 18.0 ml of cloning medium. Cloning medium must be prewarmed at 37°C.

3. Seed one 96-well tissue culture plate with 200 µl/well of the 50 cells/ml suspension, and another plate with 200 µl/well of the 5 cells/ml suspension. On the average, the plates should contain 10 and 1 cells/well, respectively.

4. Incubate the plates in a humidified, 37°C, 5% CO_2 incubator for approximately 7 to 14 days. In the meantime, check the plates under an inverted microscope in order to ascertain growth and the presence of single or multiple colonies in each well. Eventually, colonies will become visible macroscopically.

5. If adequate growth (most wells contain single colonies only) is detected in the plate seeded at 1 cell/well, use this plate for screening and selection. If not, use the plate seeded at 10 cells/well, making sure that, if possible, only those wells exhibiting a single colony are used.

6. When the colony covers approximately 10 to 20% of the surface of the well, remove 100 µl of supernatant and test for the desired specificity. Feed the wells with 100 µl of fresh cloning medium. Feed the cells at any time if the medium becomes acidic (yellow) by aspirating approximately 100 µl of supernatant and replacing it with 100 µl of fresh cloning medium.

7. Select the desired hybridomas based on the screening. Expand to a 24-well plate. Feed the cells every 2 to 3 days by removing half of the supernatant and replacing it with medium. At this point the cells can be gradually adapted to growth in culture medium (10% FCS) without the conditioned medium, by substituting culture medium for the cloning medium in the feeding schedule.

8. If questions remain about the clonality of the hybridomas selected, repeat the procedure (steps # 1 to 7) one more time.

Comments:

1. The choice of a high-quality fetal calf serum of proven hybridoma-supporting capacity (e.g., Hyclone, Sigma) is essential.

2. For preparation of conditioned medium from SP2/0-Ag14 cells, seed cells in a 75-cm² culture flask at a initial cell density of 1 x 10⁵ cells/ml, using culture medium (DMEM-10% FCS). The SP2/0-Ag14 should be routinely screened for mycoplasma contamination. Use only cells that have been proven mycoplasma-free.

3. The reason for plating cells at 10 and 1 cells/well is that usually, primary hybridomas will exhibit less than 100% plating efficiency.

4. Aliquots of cells must be frozen whenever sufficient numbers can be obtained (step # 7).

REFERENCES

1. **Fitch, F. W. and Gajewski, T. F.**, Production of T cell clones, in *Current Protocols in Immunology*, Coligan, J.,E., Kruisbeek, A. M., Margulies, D. H., Shevach, E. M., and Strober, W., Eds., Greene Publishing and Willey-Interscience, New York, 1991, 3.13.1.

2. **Gajewski, T. F., Joyce, J. and Fitch, F. W.**, Anti-proliferative effect on IFN-gamma in immune regulation. III. Differential selection of T_h1 and T_{h2} murine helper T lymphocyte clones using rIL-2 and rIFN-gamma, *J. Immunol.*, 143, 15, 1989.

3. **Shoenfeld, Y. and Isenberg, D. A.**, Eds., *Natural Antibodies. Their Physiological Role and Regulatory Significance*, CRC Press, Boca Raton, 1993.

4. **Nutman, T.B.**, Long-term and immortalized human cell lines and clones, in *Current Protocols in Immunology*, Coligan, J. E., Kruisbeek, A. M., Margulies, D. H., Shevach, E. M., and Strober, W., Eds., Green Publ. and Wiley-Interscience, New York, 1992, 7.19.1.

5. **Gosselin, E. J., Wardwell, K., Gosselin, D. R., Alter, N., Fisher, J. L. and Guyre, P. M.**, Enhanced antigen presentation using human Fcγ receptor (monocyte/macrophage) - specific immunogens, *J. Immunol.*, 149, 3477, 1992.

6. **Lanzavecchia, A.**, Antigen-specific interaction between T and B cells, *Nature*, 314, 537, 1985.

7. **Goding, J. W.**, Antibody production by hybridomas, *J. Immunol. Methods*, 39, 285, 1980.

8. **Yokoyama, W. M.**, Production of monoclonal antibodies, in *Current Protocols in Immunology*, Coligan, J. E., Kruisbeek, A. M., Margulies, D. H., Shevach, E. M., and Strober, W., Eds., Green Publ. and Wiley-Interscience, New York, 1992, 2.5.1.

9. **Parks, D. R., Bryan, V. M., Oi, V. T., and Herzenberg, L. A.**, Antigen-specific identification and cloning of hybridomas with a fluorescence-activated cell sorter, *Proc. Natl. Acad. Sci. U.S.A.*, 76, 1962, 1979.

Chapter 5

CYTOTOXIC CELLS

I. CELL-MEDIATED CYTOTOXICITY

Cell-mediated cytotoxicity is a general term including three different mechanisms: antibody-dependent cell-mediated cytotoxicity (ADCC), natural killer (NK)/lymphokine-activated killer (LAK) cell cytotoxicity and T cell cytotoxicity. In an ADCC cytotoxic reaction, effector cells mediate target cell lysis by binding their Fc receptors to the Fc fragment of antibodies bound to target cells. The other two mechanisms occur without a need for antibodies.

Several different ways of target cell labeling for the assessment of cytotoxic activity have been used. The two most common techniques use ^{51}Cr release[1-3] from ^{51}Cr-labeled targets or an enzymatic assay measuring release of intracellular substances. All these methods are based on measuring cell death judged from plasma membrane disintegration and the consequent release of cytoplasm. DNA fragmentation, a common event occurring during apoptosis, is also an early event in the cell death caused by cytotoxic cells.[4] The JAM test[5] is a sensitive and easy test for DNA fragmentation and cell death.

A. ADCC[6]

Materials and Reagents

RPMI 1640 medium with 25 mM HEPES and 10% FCS
5% (v/v) Triton X-100 (Sigma)
6-ml plastic centrifuge tubes
96-well round-bottom microtiter plates with cover (Costar)
15-ml conical centrifuge tubes
Tubes suitable for your γ-counter
γ-counter
75-cm^2 tissue culture flask (Costar)
^{51}Cr-labeled target cells (see Section D)
Ficoll-Hypaque-separated peripheral blood cells (see Chapter 1) or purified
    NK cells
Humidified 37°C, 5% CO_2 incubator

67

Protocol:

1. Dilute effector cells to 1 x 10^7/ml in medium and antibody-coated and non-antibody-coated ^{51}Cr-labeled target cells to 1 x 10^5/ml in medium.
2. Prepare 0.8 ml each of 1:2, 1:4 and 1:8 dilutions of effector cells in medium using 6-ml centrifuge tubes.
3. Pipettete 0.1 ml of undiluted effector cells from left to right along a row of a 96-well round-bottom microtiter plate for each assay condition. Make a duplicate row in the same fashion. Repeat for the 1:2, 1:4, and 1:8 dilutions. Leave one row empty for each condition.
4. Add 0.1 ml antibody-coated ^{51}Cr-labeled target cells to the first row and 0.1 ml non-antibody-coated ^{51}Cr-labeled target cells to the second row for each condition. Add 0.1 ml antibody-coated ^{51}Cr-labeled target cells into the first 6 wells of an empty row, and 0.1 ml non-antibody-coated ^{51}Cr-labeled target cells into the other 6 wells.
5. Add 0.1 ml of 5% Triton X-100 to three of the wells containing antibody-coated target cells and non-antibody-coated target cells. Add 0.1 ml of medium into remaining control wells.
6. Centrifuge the plates for 3 minutes at 100 x g at room temperature.
7. Incubate for 18 hours in a humidified 37°C, 5% CO_2 incubator.
8. Centrifuge the plates for 5 minutes at 550 x g at room temperature. Carefully transfer the supernatants into small tubes optimal for your γ-counter.

B. CytoTox 96™ CYTOTOXICITY ASSAY

This method is a colorimetric nonradioactive alternative to the traditional ^{51}Cr-release-based cytotoxicity techniques. The assay measures lactate dehydrogenase release from lysed cells. Besides easier handling, mostly because of it being nonradioactive, another advantage of this technique is its much longer expiration time compared to the rather short half-life of ^{51}Cr (27.7 days). Therefore, it is particularly useful to use CytoTox 96™ when you perform cytotoxicity experiments only occasionally. The sensitivity of this assay is fully comparable to ^{51}Cr-release assay,[7,8] however, this technique cannot be used for evaluation of erythrocyte lysis.

Materials and Reagents

Isolated NK cells
Target cells
RPMI 1640 medium

CytoTox 96™ cytotoxicity assay (Promega)
V- or round-bottom 96-well plates (Costar)
ELISA reader equipped with 490-nm filter

Protocol:

1. Add target and effector cells to wells according to your experimental protocol.
2. Add cells for spontaneous release control, maximum release control and effector cell release control.
3. Add culture medium and lysis solution for volume correction control.
4. Add culture medium for culture medium background.
5. Centrifuge plates at 250 x g for 4 minutes at room temperature.
6. Incubate plates for 4 hours at 37°C.
7. Add lysis solution (10 µl per 100 µl of supernatant) 45 minutes prior to further centrifugation.
8. Centrifuge plates at 250 x g for 4 minutes at room temperature.
9. Transfer 50 µl of supernatant from each well into a new round-bottom 96-well plate.
10. Add 50 µl reconstituted substrate mix (12 ml of assay buffer added to a bottle of substrate mix; once resuspended, protect from light and store at -20°C for up to 2 months) to each well with the transferred supernatant.
11. Incubate plates for 30 minutes at room temperature.
12. Add 50 µl of stop solution and read absorbance at 490 nm.

Comments:

1. Volume correction control adjusts for the volume changes caused by addition of the lysis solution. Culture medium background corrects for LDH activity caused by the level of animal sera and by the presence of phenol red in medium. You can skip this step by using a phenol red-free medium and by lowering the amount of serum.
2. As the content of lactate dehydrogenase in various target cells varies, preliminary experiments for optimization of their number are recommended.

C. LAK/NK CYTOTOXICITY

Materials and Reagents

Effector cells
RPMI 1640 medium with 25 mM HEPES and 10% FCS
5% (v/v) Triton X-100
6-ml plastic centrifuge tubes
96-well round-bottom microtiter plates with cover (Costar)
15-ml conical centrifuge tubes
Tubes suitable for your γ-counter
γ-counter
^{51}Cr-labeled target cells (see Section D)
2% SDS
Ficoll-Hypaque-separated peripheral blood cells (see Chapter 1) or purified NK cells
Humidified 37°C, 5% CO_2 incubator

Protocol:

1. Dilute effector cells to appropriate concentration depending on the experimental design. It is recommended to use three different effector: target ratios. The actual ratio used is dependent on factors such as target sensitivity and the effector cell potency.
2. Add 100 µl of effector cells into wells of the 96-well round-bottom microtiter plates in triplicate. Leave row A blank.
3. Add 100 µl of medium into the first 3 wells of row A for spontaneous ^{51}Cr release control.
4. Add 50 µl of ^{51}Cr-labeled target cells to each well containing effector cells and to the second three wells of a row A. These three wells will serve as a maximum lysis control.
5. Cover the plate and centrifuge it at 50 x g for 5 minutes at room temperature with brake off.
6. Incubate the plate for 4 hours in a humidified 37°C, 5% CO_2 incubator.
7. Add 100 µl of 2% SDS into those wells for maximum lysis control (second three wells, row A).
8. Transfer 50 µl of supernatant into the counting tubes and measure the radioactivity of the samples in a γ-counter.

Calculation:

$$\% \text{ lysis } = \frac{\text{experimental cpm - spontaneous cpm}}{\text{maximum cpm - spontaneous cpm}} \times 100$$

D. ^{51}Cr LABELING

Materials and Reagents

Target cells
$Na_2^{51}CrO_4$, 5 mCi/ml, specific activity 250 to 500 mCi/mg, (Amersham)
RPMI 1640 medium with 25 mM HEPES and 10% FCS
15-ml conical centrifuge tubes
Tubes suitable for your γ-counter
γ-counter

Protocol:

1. Resuspend cells in medium at a concentration 1 x 10^7/ml. Check the viability; use only cells with viability higher than 95%.
2. Add $Na_2^{51}CrO_4$ to the cells (500 µCi per 1 x 10^7 cells) and incubate for 60 minutes at 37°C with occasional shaking.
3. Wash the cells at least five times by centrifugation at 300 x g for 10 minutes at room temperature. Check the radioactivity in the supernatant from last centrifugation. If still significantly higher than a background, repeat the washing procedure.

Comments:

1. A major disadvantage of this very common technique is the rather high spontaneous ^{51}Cr release from target cells, especially when the cells are grown in monolayers and need to be trypsinized in order to obtain a single cell suspension prior to labeling.

E. ISOLATION OF NK CELLS

Materials and Reagents

Blood

71

15-ml conical centrifuge tubes
RPMI 1640 medium with 10% FCS
Materials for isolation of mononuclear cells, panning, nylon wool column
  Percoll separation (see Chapter 1)
15-ml glass centrifuge tubes

TABLE 1

Discontinuous Percoll Density Gradient

Fraction	Percoll* (ml)	Density (g/ml)	Volume (ml)**
1	2.55	1.053	2.5
2	2.70	1.060	2.5
3	2.85	1.063	2.5
4	3.00	1.068	1.5
5	3.15	1.073	1.5
6	3.30	1.077	1.5
7	4.00	1.080	1.5

*Stock solution

**Volume of each fraction in a 15-ml glass centrifuge tube

Protocol:

1. Isolate mononuclear cells from peripheral blood (see Chapter 1).
2. After five washes in medium (250 x g for 5 minutes at 4°C), deplete the suspension of monocytes and some B cells by panning (see Chapter 1).
3. After two washes in medium (250 x g for 5 minutes at 4°C), pass the cells through a nylon wool column to remove T cells.
4. After two additional washes in medium (250 x g for 5 minutes at 4°C), fractionate cells on a discontinuous Percoll gradient[9] (see Table 1).
5. Carefully layer 5 to 10 x 10⁶ cells on top of the gradient and centrifuge at 550 x g for 30 minutes at room temperature with brakes off.
6. Collect cells from resulting bands and wash them twice with RPMI 1640 medium by centrifugation at 300 x g for 5 minutes at room temperature. NK cells are usually present in the first two bands from the top between the low-density fractions.[9]

Comments:

1. Individual steps can be substituted with the negative selection using magnetic microspheres coupled with appropriate antibodies (see Chapter 1). Another possibility is to use a positive selection using magnetic microspheres coupled with CD56 and CD16 antibodies.

F. GENERATION OF LAK CELLS

Materials and Reagents

Isolated NK cells
RPMI 1640 medium with 10% FCS
Human rIL-2
25-cm^2 tissue culture flasks

Protocol:

1. Resuspend isolated NK cells (1 x 10^6/ml) in RPMI 1640 medium containing 1000 U/ml rIL-2.
2. Put 10 ml of cell suspension into 25-cm^2 tissue culture flasks and incubate for 3 days in a humidified 37°C, 5% CO$_2$ incubator.
3. Shake the flasks well to remove all adherent cells. Rinse the flasks with RPMI 1640 medium. Centrifuge LAK cells twice in RPMI 1640 medium without rIL-2 at 500 x g for 5 minutes at room temperature. Count the cells.

G. PROLIFERATION OF NK CELLS[10]

Materials and Reagents

Isolated NK cells
Peripheral blood mononuclear cells (see Chapter 1)
RPMI 1640 medium with 2.5% human AB serum
RPMI 1640 medium with 15% human AB serum
Ionomycin (Calbiochem)
25-cm^2 and 150-cm^2 tissue culture flasks
Daudi cells (ATCC; CCL-213)
PMA (Sigma)
PHA (Sigma)
Humidified 37°C, 5% CO$_2$ incubator

Protocol:

1. Prepare leukocyte conditioned medium (LCM): incubate 4 x 10⁸ isolated peripheral blood mononuclear cells from each of two donors with 2.5 x 10⁸ irradiated (8,000 rad) Daudi cells in 500 ml of RPMI 1640 medium with 2.5% human AB serum containing 10 ng/ml PMA and 4 µg/ml PHA in a 150-cm² tissue culture flasks for 120 minutes at 37°C. Wash the cells three times in medium and resuspend in RPMI 1640 medium with 2.5% human AB serum. Incubate in the same flask for 48 hours at 37°C and collect the cell-free supernatant.

2. Incubate 1 x 10⁶ purified NK cells/ml in 25-cm² tissue culture flasks in RPMI 1640 medium with 15% human AB serum, 20% LCM, and 750 ng/ml ionomycin for 24 hours in a humidified 37°C, 5% CO_2 incubator. Resuspend the cells in RPMI 1640 medium with 15% human AB serum and 20% LCM and incubate in the same flask for 10 to 14 days. Add fresh 20 % LCM to keep concentration of 1 to 2 x 10⁶ cells/ml.

Comments:

1. LCM can be passed through a 0.45-µm filter and stored at -70°C.
2. It is important not to incubate cells in medium supplemented with ionomycin for more than 24 hours, as exposure to ionomycin for longer time produces inferior proliferation of NK cells.

H. JAM TEST[5]

Materials and Reagents

Isolated NK cells
RPMI 1640 medium with 5 % FCS
96-well flat- or round-bottom tissue culture plates (Costar)
24-well tissue culture plates (Costar)
Scintillation vials
Glass fiber filters
96-well harvester
[³H-methyl] thymidine (Amersham)
Liquid scintillation counter
Humidified 37°C, 5% CO_2 incubator

74

Protocol:

1. Label target cells with [³H-methyl] thymidine as follows: Add [³H-methyl] thymidine to a final concentration of 2.5 to 5 μCi/ml (9.25 to 18.5 x 10⁴ Bq/ml) to growing target cells (concentration 2.5 to 5 x 10⁵ cells/ml) in 24-well tissue culture plates and incubate for 4 to 6 hours. Wash the cells once in RPMI 1640 medium with 5% FCS and resuspend to 1 x 10⁵ cells/ml.

2. Dilute effector cells to an appropriate concentration depending on the experimental design. It is recommended to use three different effector: target ratios. The actual ratio used is dependent on factors such as target sensitivity and the effector cell potency.

3. Add 100 μl of effector cells into wells of the 96-well plates in triplicate. Leave row A empty.

4. Add 100 μl of medium into the first 3 wells of row A for spontaneous death control.

5. Add 100 μl of [³H-methyl] thymidine-labeled target cells to each well containing effector cells and to the second 3 wells of a row A (maximum lysis control).

6. Incubate the plate for 2 hours in a humidified 37°C, 5% CO_2 incubator.

7. Add 100 μl of 2% SDS into those wells for maximum lysis control (second three wells, row A).

8. Aspirate the cells and their medium on to fiber glass filters using a cell harvester (normally used for harvesting cells for proliferation assay; see Chapter 3), wash the filters, dry, add liquid scintillation fluid and count in a liquid scintillation counter.

Calculation:

$$\% \text{ specific DNA loss} = \frac{(T-E)-(T-S)}{T-(T-S)} \times 100$$

E = experimentally retained DNA in the presence of NK cells; S = retained DNA in the absence of NK cells (spontaneous), and T = total incorporated label (maximum).

REFERENCES

1. Sanderson, A. R., Cytotoxic reaction of mouse isoantisera: preliminary consideration, *Br. J. Exp. Pathol.*, 45, 398, 1964.
2. Brunner, K. T., Mauel, J., Cerottini, J. C. and Chapuis, B., Quantitative assay of the lytic action of immune lymphoid cells on ^{51}Cr-labelled allogeneic target cells in vitro; inhibition by isoantibody and by drugs, *Immunology*, 14, 181, 1968.
3. Gorer, P. A. and O'Gorman, P., The cytotoxic activity of isoantibodies in mice, *Transpl. Bull.*, 3, 142, 1956.
4. Duke, R. C., Chervenak, R. and Cohen, J. J., Endogenous endonuclease-induced DNA fragmentation: an early event in cell-mediated cytolysis, *Proc. Natl. Acad. Sci. U.S.A.*, 80, 6361, 1983.
5. Matzinger, P. The JAM test. A simple assay for DNA fragmentation and cell death, *J. Immunol. Methods*, 145, 185, 1991.
6. Nelson, D. L., Kurman, C. C. and Serbousek, D. E., ^{51}Cr release assay of antibody-dependent cell-mediated cytotoxicity (ADCC), in *Current Protocols in Immunology*, Colignan, J. E., Kruisbeek, A. M., Margulies, D. H., Shevach, E. M. and Strober, W., Eds., Greene Publishing and Wiley-Interscience, New York, 1991, 7.27.1.
7. Korzeniewski, C. and Callewaert, D. M., An enzyme-release assay for natural cytotoxicity, *J. Immunol. Methods*, 64, 313, 1983.
8. Decker, T. and Lohmann-Matthes, M L., A quick and simple method for the quantitation of lactate dehydrogenase release in measurements of cellular cytotoxicity and tumor necrosis factor (TNF) activity, *J. Immunol. Methods*, 115, 61, 1988.
9. Timonen, T., Reynolds, C. W., Ortaldo, J. R. and Herberman, R. B., Isolation of human and rat natural killer cells, *J. Immunol. Methods*, 51, 269, 1982.
10. Robertson, M. J., Manley, T. J., Donahue, C., Levine, H. and Ritz, J., Costimulatory signals are required for optimal proliferation of human natural killer cells, *J. Immunol.*, 150, 1705, 1993.

Chapter 6

ASSAYS OF CYTOKINES

I. INTRODUCTION

Cytokines are a group of small glycoproteins (~15 to 45 kDa) whose main function is to act as intercellular signals. Cytokines are secreted by and act on lymphocytes, monocyte/macrophages and a variety of other cells, including cells of nonhemopoietic origin. Although many physiologic processes are regulated by cytokines, the regulation of inflammatory and immunologic responses are two of their main functions.[1] To date, more than 80 different cytokines have been identified and their genes cloned.[2]

Generally, the synthesis of cytokines is not constitutive, but inducible by a variety of stimuli, depending on the nature of the secreting cell. In the case of lymphocytes, immunological stimulation by antigens or mitogens is the main trigger of cytokine secretion, whereas endotoxin and other microbial components are the main substances that induce cytokine synthesis and release from monocytes and other inflammatory cells.[3] Thus, since cytokine synthesis is a consequence of cellular activation, the *in vitro* or *in vivo* assessment of cytokine synthesis has been commonly used as a parameter of cellular activation. In addition, since different cell types and even different cell subsets secrete different sets of cytokines (i.e., Th1 and Th2 subsets of CD4+ T cells),[4,5] identification and quantitation of different cytokines allow the study of the particular cell types or subsets activated by a particular signal or during a particular pathologic process.

Although the measurement of cytokines is well established at the levels of basic and experimental immunology, evidence has indicated that cytokine assays may also have significant clinical value, particularly for prognostic or monitoring purposes.[6] Whether for research or clinical purposes, cytokine expression, synthesis and secretion can be investigated by a variety of different techniques, including:

1. Assay of secreted cytokines in supernatants or biological fluids.
2. Analysis of expression of cytokine-specific mRNA by Northern blotting or reverse-transcriptase polymerase chain reaction (RT-PCR).
3. Detection of cytokines in tissue samples by immunohistochemistry or *in-situ* hybridization.

The subject of this chapter will be the assay of secreted cytokines.

77

A. ASSAY OF SECRETED CYTOKINES

The concentration of different cytokines present in biological fluids (e.g., serum, plasma, urine, saliva) or in tissue culture supernatants can be determined based on their activity (Section II) or on their immunogenicity or receptor-binding properties (Section III). These two different techniques have inherent advantages and disadvantages that need to be taken in consideration depending on the purpose of the cytokine measurement. Both types of assays are discussed below.

B. STANDARDIZATION OF CYTOKINE MEASUREMENTS

All cytokine assays, whether bioassays or immunoassays require the use of standard preparations of recombinant cytokines of known activity and concentration. Although many human and murine cytokines are available from commercial sources and these preparations can be routinely used as standards, cytokine assays should be calibrated against reference cytokine preparations. Reference standards are available from the Biological Response Modifiers Program (BRMP), National Cancer Institute (Frederick, MD), and the Division of Microbiology and Infectious Diseases, National Institute for Allergy and Infectious Diseases (NIAID, Bethesda, MD) in the U.S.; and from the National Institute for Biological Standards and Control (NIBSC) in London, Great Britain. These reagents are provided in small amounts for calibration or standard purposes only, and they are defined in terms of units of biological activity (U/ml). Although many commercially available cytokine preparations and ELISA kits are based on cytokine concentration expressed in terms of weight units/ml (e.g., ng/ml), reference standards can be used to calibrate the specific activity (e.g., Units/ng) of such preparations.

II. BIOASSAYS

Before the availability of large amounts of recombinant cytokines and monoclonal anti-cytokine antibodies, cytokines were identified and assayed based on their biologic activity, normally using "indicator cells", which could be either freshly isolated cells or established cell lines. Although bioassays are still a convenient way to identify and/or quantitate cytokines, it has become evident that most "indicator cells" are seldom cytokine specific, and can usually respond to other cytokines in addition to the one they are intended to measure.[6] Hence, specificity may be a problem. For example, HT-2 or CTLL cell lines, which had been extensively used for the measurement of IL-2,[7,8] were later found to respond also to IL-4,[9] and more recently to IL-15.[10] Nevertheless, when used in conjunction with appropriate neutralizing anti-cytokine monoclonal antibodies to either confirm the identity of the cytokine being measured or to neutralize potentially interfering cytokines, bioassays still offer relatively simple and sensitive means for cytokine assay. The main advantages and disadvantages of bioassays are given below.

Advantages:

1. Measure only "functional" cytokines (true reflection of biological activity);
2. Very sensitive (~ 10 pg/ml);
3. Relatively inexpensive.

Disadvantages:

1. Indicator cells are usually not cytokine specific (may respond to more than one cytokine and may be susceptible to potentiating or inhibitory effects of other cytokines);
2. Susceptible to inhibitors (e.g., cytokine antagonists, soluble cytokine receptors) present in biological fluids;
3. Do not differentiate between subtypes of cytokines (e.g., IL-1α and IL-1β, TNFα and TNFβ, etc.);
4. Reproducibility may be compromised due to variability in indicator cells;
5. Lengthy (results may take from one to several days).

Although there could be as many different bioassays as different cytokines, most are based on:

a) Proliferation: Measurement of proliferative response of indicator cell lines in response to the appropriate cytokine. Usually performed by assessing tritiated thymidine uptake or MTT-reduction (see Chapter 3).

b) Cytotoxicity: Based on the cytotoxicity of certain cytokines for appropriate indicator cell lines or on protection from viral infection and lysis. Usually performed by assessing inhibition of cellular proliferation, ^{51}Cr release or decreased cell viability.

c) Differentiation: Based on the induction of a differentiation function on a target cell. These could include expression of surface molecules, including MHC class I and II molecules, secretion of immunoglobulins, including different isotypes, formation of colonies on soft agar, etc.

79

In conclusion, cytokine bioassays can provide a very useful way of measuring cytokines if adequate precautions are taken to insure their specificity (i.e., through neutralizing anti-cytokine monoclonal antibodies). Bioassays for the most commonly assayed cytokines are described below. Note that even though most of the protocols for bioassays based on proliferation assays are performed by tritiated thymidine uptake, these can also be adapted to an MTT-reduction assay (see Chapter 3). Immunoassays will be described later in this chapter.

A. INTERLEUKIN-1

Interleukin-1 (IL-1) is an extremely pleiotropic cytokine secreted by cells of the monocyte/macrophage lineage and many other types of cells of lymphocytic, myeloid and nonhemopoietic origin.[11] IL-1 plays a key role as a mediator of inflammation and has important activities promoting activation and differentiation of T and B lymphocytes.[2,11] In humans and mice, IL-1 exists in two different forms, IL-1α and IL-1β, encoded by different genes sharing only moderate sequence homology.[2,11] Both IL-1α and IL-1β are synthesized as large cytoplasmic precursors which are proteolytically processed and released as the mature forms.[2,11] In addition to secreted IL-1, membrane-bound IL-1 activity has been reported in fixed cells.[12] Two different receptors for IL-1 (type I and type II IL-1R) have been identified.[13,14] Moreover, the activity of IL-1 *in vivo* seems to be regulated by an endogenous IL-1 receptor antagonist (IL-1Ra), a molecule that exhibits approximately 26% sequence homology with IL-1β, and that is able to bind to IL-1Rs without triggering any biologic activity.[15]

Many different bioassays for IL-1 have been described, including the co-stimulatory activity on thymocytes[16] and the D10.G4.1 T cell line,[17] induction of IL-2 secretion from the lymphoma cell lines, LRBM.33.1A5[18] or EL-4,[19] proliferation of fibroblasts,[20] and inhibition of the proliferation of the A375 human malignant melanoma cell line.[21] The first two assays will be described in this section. Because human IL-1 is active on murine systems, both human and murine IL-1 can be assayed using either protocol.

1. Thymocyte Co-Stimulation Assay

This assay is based on the ability of IL-1 to potentiate the proliferation of murine thymocytes in response to phytohemagglutinin (PHA). This assay does not differentiate between IL-1α and IL-1β activities.

Materials and Reagents

Mouse (C3H/HeJ) thymocyte suspension
Culture medium: RPMI-1640 containing 10% heat-inactivated FCS and
    supplemented with 2 mM L-glutamine, 10 mM HEPES, 100 U/ml
    penicillin, 100 µg/ml streptomycin, 50 µM 2-mercaptoethanol and 1 mM
    sodium pyruvate
Washing buffer: PBS containing 1% FCS (filter-sterilized)
Phytohemagglutinin-P (PHA, Sigma Cat. No. L-9132)
Recombinant human or mouse IL-1 standard (e.g., Genzyme, Cistron)
[^{3}H-methyl] thymidine
96-well tissue culture plates (flat bottom), sterile
15- and 50-ml centrifuge tubes
60 x 15-mm Petri dishes
Pasteur pipettes
Refrigerated centrifuge (set at 4°C)
CO_2 incubator (humidified, set at 37°C and 5% CO_2)
Cell harvester (e.g., Skatron)
Glass fiber filter mats (Skatron)
Scintillation vials
Scintillation cocktail (e.g., Econo-Safe™, Research Products International)
Scintillation counter

Protocol:

(Note: All materials and reagents must be sterile and proper aseptic technique
must be used when handling the cells.)

1.   Sacrifice the animal and prepare a thymocyte suspension by removing
the thymus, rinsing it in a 60 x 15 mm-Petri dish with 10 ml washing
buffer, and then grinding it with the aid of two sterile frosted glass
slides in another sterile Petri dish containing 10 ml of ice-cold washing
buffer.

2.   Collect the cell suspension and transfer it to a 50-ml centrifuge tube
with a sterile Pasteur pipette. Wash the cells once by resuspending them
in 50 ml washing buffer and centrifuging for 10 minutes at 200 x g in
a refrigerated (4°C) centrifuge.

3.   Aspirate the supernatant and resuspend the cells in culture medium.
Count. Adjust cell density with culture medium to a cell density of 10^7
cells/ml.

4.   Determine the number of wells required. Each sample and standard
dilution should be assayed in triplicate. In addition, include a negative

control, in which the thymocytes are cultured with no added source of IL-1. (A minimum of 6 or 7 different dilutions of the IL-1 standard should be tested.)

5. Add 100 µl of the thymocyte suspension per well. Add 50 µl of a (4X) PHA solution made in culture medium. (The final concentration of PHA in the cultures should be determined beforehand. For this, first test several concentrations of PHA [0.1 to 5 µg/ml] and choose a concentration giving suboptimal proliferation and maximal enhancement by a rIL-1 standard.)

6. Add samples and rIL-1 standards (1 to 1,000 pg/ml) diluted in culture medium. Each dilution should be assayed in triplicate. Incubate for 48 hours at 37°C and 5% CO_2.

7. Add 20 µl/well of a 50-µCi/ml solution of [^{3}H-methyl] thymidine in culture medium (final concentration is 1 µCi/well) for the last 6 hours of incubation.

8. Harvest the cells onto glass fiber mats with the aid of a cell harvester. Determine [^{3}H-methyl] thymidine uptake by liquid scintillation counting (see Chapter 3).

9. Calculate IL-1 concentration. Using semilog or probit paper, plot standard IL-1 concentration on the x axis vs. proliferation (in cpm) on the y axis. Calculate sample concentration by comparison to standard curve and multiplication by dilution factor.

Comments:

1. Although thymocytes from other strains are also responsive to IL-1, thymocytes from C3H/HeJ mice are generally used in this assay due to their high responsiveness. The use of animals 6 to 12 weeks old is recommended.

2. When assaying unknown samples, test several dilutions so that a curve for those samples can be generated. Concentrations can then be calculated by comparison of sample dilution giving 50% maximal response with the standard concentration giving 50% maximal proliferation. For example:

    Sample concentration = Standard. Conc. (50% max) x Sample Dilution (50% max)

3. Several other cytokines, including IL-2, IL-6 and TNFα may also induce proliferation of thymocytes in the presence of mitogen.

2. Co-stimulation Assay Using the D10.G4.1 Cell Line

This assay is based on the ability of IL-1 to stimulate the proliferation of the T helper cell line, D10.G4.1, in the presence of a suboptimal concentration of Concanavalin A. D10.G4.1 is a murine CD4⁺ T cell clone with specificity for conalbumin presented by accessory cells expressing the I-A^k MHC class II molecule.[22] This assay does not differentiate between IL-1α and IL-1β activities.

Materials and Reagents

D10.G4.1 cells (ATCC Cat. No. TIB 224)
Concanavalin A (ConA, Sigma Cat. No. C-5275)
Culture medium: RPMI-1640 containing 10% heat-inactivated FCS and supplemented with 2 mM L-glutamine, 10 mM HEPES, 100 U/ml penicillin, 100 µg/ml streptomycin, 50 µM 2-mercaptoethanol and 1 mM sodium pyruvate
Washing buffer: PBS containing 1% FCS (filter-sterilized)
Recombinant human or mouse IL-1 standard (e.g., Genzyme, Cistron)
[³H-methyl] thymidine
96-well tissue culture plates (flat bottom), sterile
15- and 50-ml centrifuge tubes
Pasteur pipettes
Refrigerated centrifuge (set at 4°C)
CO₂ incubator (humidified, set at 37°C and 5% CO₂)
Cell harvester (e.g., Skatron)
Glass fiber filter mats (Skatron)
Scintillation vials
Scintillation cocktail (e.g., Econo-Safe™, Research Products International)
Scintillation counter

Protocol:

(Note: All materials and reagents must be sterile and proper aseptic technique must be used when handling the cells.)

1. Harvest the D10.G4.1 cells approximately 7 days after the last stimulation (see maintenance of D10.G4.1 cells). Wash the cells twice in 20 ml of ice-cold washing buffer, centrifuging at 200 x g at a temperature of 4°C. Resuspend cells in culture medium containing 2.5 µg/ml ConA to a cell density of 10⁵ cells/ml.

2. Determine the number of wells required. Each sample and standard

dilution should be assayed in triplicate. In addition, include a negative control, in which the D10.G4.1 cells are cultured with no added source of IL-1. (A minimum of 6 or 7 different dilutions of the IL-1 standard [1 to 1,000 pg/ml] should be tested.)

3. Add 100 µl of the cell suspension per well.

4. Add test samples and standards diluted in culture medium (100 µl/well). Assay each dilution in triplicate.

5. Incubate for 72 hours at 37°C and 5% CO_2. Add 20 µl/well of a 50-µCi/ml solution of [^{3}H-methyl] thymidine in culture medium (final concentration is 1 µCi/well) for the last 6 hours of incubation.

6. Harvest the cells onto glass fiber filters with the aid of a cell harvester. Determine [^{3}H-methyl] thymidine uptake by liquid scintillation counting (see Chapter 3).

9. Calculate IL-1 concentration. Using semilog or probit paper, plot standard IL-1 concentration on the x axis vs. proliferation (in cpm) on the y axis. Calculate sample concentration by comparison to standard curve and multiplication by dilution factor.

3. Maintenance of D10.G4.1 Cells

Materials and Reagents

Spleen cells from mice expressing I-A^k MHC class II (e.g., AKR, CBA/J, C3H/HeJ)
Mitomycin C (Sigma Cat. No. M-0503)
Conalbumin (Sigma Cat. No. C-0755)
Recombinant human or murine IL-2 (rIL-2) (e.g., Genzyme)
Culture medium: RPMI-1640 containing 10% heat-inactivated FCS and supplemented with 2 mM L-glutamine, 10 mM HEPES, 100 U/ml penicillin, 100 µg/ml streptomycin, 50 µM 2-mercaptoethanol and 1 mM sodium pyruvate
Washing buffer: PBS containing 1% FCS (filter-sterilized)
24-well tissue culture plates or 100 x 15 mm-Petri dishes
60 x 15-mm Petri dishes
15- and 50-ml centrifuge tubes
Pasteur pipettes
Refrigerated centrifuge (set at 4°C)
CO_2 incubator (humidified, set at 37°C and 5% CO_2)
Sterile forceps

Protocol:

(Note: All materials and reagents must be sterile and proper aseptic technique must be used when handling the cells.)

1. Sacrifice mice and aseptically remove the spleens. Wash the spleens in a 60 x 15-mm Petri dish with 10 ml ice-cold washing buffer and prepare a splenic cell suspension by gently teasing the spleens with a pair of sterile forceps. Wash the cells twice with 40 ml of cold washing buffer.

2. Resuspend the cells in culture medium to a cell density of 2 x 10^7 cells/ml. Add mitomycin C to a final concentration of 50 µg/ml and incubate in a 37°C water bath for 30 to 60 minutes with occasional agitation. Wash the cells four times with washing buffer. Count and resuspend in culture medium to a density of 1 x 10^7 cells/ml.

3. Using 24-well tissue culture plates, 100 x 15-mm Petri dishes or tissue culture flasks (depending on the number of cells and volume of medium), culture the D10.G4.1 cells at 37°C, 5% CO_2 and a final density of 1 to 2 x 10^5/ml in the presence of mitomycin C-treated splenic cells (final density of 5 x 10^6/ml), conalbumin (100 µg/ml) and rIL-2 (100 U/ml). Repeat this stimulation every two weeks. Passage the cells at a 1:5 dilution in IL-2-containing medium (100 U/ml) if the culture medium becomes yellow.

Comments:

1. IL-2 may mimic IL-1 in this assay.

2. Conditioned medium from ConA-activated splenocytes can be used as a source of IL-2 (instead of the rIL-2) for the maintenance of the D10.G4.1 cells. To prepare the conditioned medium, prepare a splenic cell suspension from BALB/c mice and culture at 10^7 cells/ml in the presence of ConA (5 µg/ml) for 24 hours at 37°C and 5% CO_2. Collect the supernatant by centrifugation, filter and store at -20°C until use. Use at 5 to 10% final concentration.

B. INTERLEUKIN-2

Interleukin-2 (IL-2) is a cytokine produced by activated CD4[+] and some CD8[+] T lymphocytes.[4,23] In addition to being the major T cell growth factor for both human and murine T cells, IL-2 also stimulates growth and differentiation of cytotoxic T cell precursors (CTLs), growth and cytotoxic activity of NK cells, proliferation and differentiation of activated human B cells and activation of monocytes.[2,23] The biological effects of IL-2 are mediated by binding to high-affinity IL-2 receptors (IL-2Rs), which are composed of at least three different subunits: IL-2Rα (CD25, p55, Tac antigen), IL-2Rβ (p75) and IL-2Rγ.[24] Signaling through the IL-2R requires the IL-2Rγ chain, which is also shared with other cytokine receptors, including IL-4R, IL-7R, and IL-13R.[25-28] Upon activation, T cells also release truncated forms of the α-subunit, which appear as soluble IL-2Rs in supernatants or biological fluids.[29]

The bioassays for IL-2 are all based on its ability to stimulate the proliferation of murine T cell lines, such as HT-2 and CTLL-2.[7,8] These cell lines respond to and can be used to assay both human and murine IL-2. However, since murine IL-4 also stimulates the proliferation of both HT-2 and CTLL-2 cells,[9] assays involving mouse-derived samples need to be performed in the presence of neutralizing anti-mouse IL-4 antibodies in order to make the assay monospecific for IL-2. Conversely, performing the proliferation assay in the presence of neutralizing anti-mouse IL-2 antibodies, makes the assay specific for IL-4. Human IL-4 is not active in murine systems, and thus no blocking anti-human IL-4 antibodies are required for the human IL-2 assay.

1. HT-2 or CTLL-2 Proliferation Assay for IL-2/IL-4

This assay, which is routinely used to measure IL-2 and/or IL-4, is based on the proliferation of the IL-2/IL-4 -responsive T cell lines, HT-2[7] or CTLL-2.[8]

Materials and Reagents

Indicator cells: HT-2 (ATCC Cat. No. CRL 1841) or CTLL-2 cells (ATCC Cat. No. TIB 214)
Recombinant human or mouse IL-2 standard
Recombinant mouse IL-4 standard
Rat monoclonal anti-mouse IL-4 (11B11) (ATCC Cat. No. HB 188)
Rat monoclonal anti-mouse IL-2 (S4B6) (ATCC Cat. No. HB 8794)
Recombinant human or murine IL-2 (rIL-2) (e.g., Genzyme)
Culture medium: RPMI-1640 containing 10% heat-inactivated FCS and
    supplemented with 2 mM L-glutamine, 10 mM HEPES, 100 U/ml
    penicillin, 100 µg/ml streptomycin, 50 µM 2-mercaptoethanol and 1 mM
    sodium pyruvate
Washing buffer: PBS containing 1% FCS (filter-sterilized)
[^{3}H-methyl] thymidine
96-well tissue culture plates (flat bottom)

86

15- and 50-ml centrifuge tubes
Pasteur pipettes
Refrigerated centrifuge (set at 4°C)
CO_2 incubator (humidified, set at 37°C and 5% CO_2)
Cell harvester (e.g., Skatron)
Glass fiber filter mats (Skatron)
Scintillation vials
Scintillation cocktail (e.g., Econo-Safe™, Research Products International)
Scintillation counter

Protocol:

(Note: All materials and reagents must be sterile and proper aseptic technique must be used when handling the cells.)

1. Harvest the indicator cells, usually at the end of the second or third day of culture (see maintenance of HT-2 and CTLL-2 cells below), wash twice with 20 ml of cold washing buffer and once with complete medium in order to remove any IL-2. Determine viability by trypan blue exclusion and count. Resuspend the cells in culture medium to a cell density of 2.5 x 10^5 cells/ml. Maintain the cells in an ice bath until use.

2. Dilute samples and standards (10 to 10,000 pg/ml) in culture medium. Assay each dilution in triplicate. Include a background proliferation control in which the indicator cells are cultured in the presence of medium alone. Normally, 6 to 7 different dilutions of the standard are required for the standard curve.

3. Add 20 to 50 µl of sample or recombinant IL-2/IL-4 standard per well.

4. If it is unknown whether samples contain IL-2, IL-4 or both, neutralizing anti-IL-2 or anti-IL-4 antibodies need to be included. Do one set of samples without antibodies, another with anti-IL-2 monoclonal antibody (S4B6, usually at a final concentration of 10 µg/ml) and another with anti-IL-4 monoclonal antibody (11B11, again at 10 µg/ml). Preincubate samples and antibodies for 30 minutes at room temperature.

5. Add enough culture medium to bring the volume of each well to 80 µl.

6. Add 20 µl of the indicator cell suspension to each well (5,000 cells/well). The final volume is 100 µl. Incubate for 24 hours at 37°C in a 5% CO_2 atmosphere.

7. Add 20 µl/well of a 50 µCi/ml solution of [^{3}H-methyl] thymidine in culture medium (final concentration is 1 µCi/well). Incubate overnight at 37°C and 5% CO_2.

8. Harvest the cells onto glass fiber mats with the aid of cell harvester. Determine [^{3}H-methyl] thymidine uptake by liquid scintillation counting (see Chapter 3).

9. Calculate IL-2 (wells containing anti-IL-4 antibodies) and/or IL-4 concentrations (wells containing anti-IL-2 antibodies) by reference to recombinant IL-2 and IL-4 standards, respectively. Using semilog or probit paper, plot standard IL-2/4 concentration on the x axis vs. proliferation (in cpm) on the y axis. Calculate sample concentration by comparison to standard curve and multiplication by dilution factor.

2. Maintenance of HT-2 or CTLL-2 Cells

Materials and Reagents

HT-2 (ATCC Cat. CRL 1841) or CTLL-2 cells (ATCC Cat. No. TIB 214)
Culture medium: RPMI-1640 containing 10% heat-inactivated FCS and supplemented with 2 mM L-glutamine, 10 mM HEPES, 100 U/ml penicillin, 100 µg/ml streptomycin, 50 µM 2-mercaptoethanol and 1 mM sodium pyruvate
Recombinant IL-2 (human or mouse)
25-cm^2 culture flasks
15- and 50-ml centrifuge tubes
Pasteur pipettes
Refrigerated centrifuge (set at 4°C)
CO$_2$ incubator (humidified, set at 37°C and 5% CO$_2$)

Protocol:

(Note: All materials and reagents must be sterile and proper aseptic technique must be used when handling the cells.)

1. Seed HT-2 or CTLL-2 cells at an initial density of 1 x 10^5 cells/ml in culture medium containing rIL-2 (10 to 20 U/ml).
2. Culture at 37°C and 5% CO$_2$ for three days. At this point the cells can be harvested and used in the IL-2/IL-4 assay.
3. Subculture every third day.

Comments:

1. It is recommended to use only indicator cells with > 90% viability after harvesting.

2. Even though these cell lines proliferate in response to murine IL-4, they cannot be maintained for long periods of time exclusively in IL-4.
3. CTLL-2 cells also proliferate in response to the newly described cytokine, IL-15.[10]
4. Conditioned medium from ConA-activated splenocytes can also be used as a source of IL-2 for maintenance of HT-2 or CTLL-2 cell lines (refer to Section A.3 in this chapter for preparation).

C. INTERLEUKIN-3

Interleukin-3 (IL-3), also known as multi-colony stimulating factor (Multi-CSF) or mast cell growth factor, is a cytokine produced primarily by activated T lymphocytes, but also by myelomonocytic cell lines (WEHI-3) and activated mast cells and NK cells.[4,30,31] IL-3 is a member or a group of cytokines whose main function is to promote the growth and differentiation of hemopoietic precursor cells. In particular, IL-3 acts on pluripotent bone marrow stem cells to induce the generation of all types of hemopoietic colonies (granulocytes, monocytes, megakaryocytes, mast cells, eosinophils) in soft agar.[32] In addition, IL-3 stimulates the growth of mast cell lines and potentiates the activities of eosinophils, basophils and mast cells.[33] The effects of IL-3 are mediated through binding to high affinity IL-3 receptors (IL-3Rs), which are heterodimers formed by an α-chain (70 kDa) that binds IL-3, and by a signal-transducing β-subunit (130 kDa) that is shared among IL-3, GM-CSF and IL-5 receptors.[34,35]

Although IL-3 and other colony-stimulating factors, such as GM-CSF, G-CSF and M-CSF can be assayed based on their activity inducing the formation of hemopoietic colonies in soft agar,[32] the availability of IL-3-dependent mast cell and myeloid cell lines, including FDCP-2,[36] FD.C/1,[37] 32D,[38] and MC/9,[39] has made possible to assay IL-3 based on simple proliferation assays, such as those already described for IL-2 and IL-4. Since human and murine IL-3 are species specific in their actions, the indicator cell lines mentioned above (all of murine origin) are suitable only for the assay of mouse IL-3. However, human IL-3 can be assayed using the IL-3-responsive erythroleukemia cell line, TF-1.[40]

1. IL-3-Dependent Cell Line Proliferation Assay

Materials and Reagents

IL-3-dependent cell line: MC/9 (mouse, ATCC Cat. No. CRL-8306); TF-1 (human, ATCC Cat. No. CRL-2003)
Recombinant IL-3 standard (mouse or human)

Culture medium: RPMI-1640 containing 10% heat-inactivated FCS and supplemented with 2 mM L-glutamine, 10 mM HEPES, 100 U/ml penicillin, 100 µg/ml streptomycin, 50 µM 2-mercaptoethanol and 1 mM sodium pyruvate
Washing buffer: PBS containing 1% FCS (filter-sterilized)
[^{3}H-methyl] thymidine
96-well tissue culture plates (flat bottom)
15- and 50-ml centrifuge tubes
Pasteur pipettes
Refrigerated centrifuge (set at 4°C)
CO_2 incubator (humidified, set at 37°C and 5% CO_2)
Cell harvester (e.g., Skatron)
Glass fiber filter mats (Skatron)
Scintillation vials
Scintillation cocktail (e.g., Econo-Safe™, Research Products International)
Scintillation counter

Protocol:

(Note: All materials and reagents must be sterile and proper aseptic technique must be used when handling the cells.)

1. Harvest indicator cells, wash twice with cold washing buffer and once with culture medium to remove any IL-3. Count and resuspend the cells to a density of 5 x 10^4 cells/ml in culture medium.
2. Dilute test samples and standards (0.1 to 100 U/ml) in culture medium. Include a negative control (background proliferation) containing medium only. Plate at 100 µl per well (in triplicate).
3. Add 100 µl of the indicator cell suspension to each well (final: 5,000 cells/well).
4. Incubate the plate for 24 hours at 37°C and 5% CO_2.
5. Add 20 µl/well of a 50-µCi/ml solution of [^{3}H-methyl] thymidine in culture medium (final concentration is 1 µCi/well). Incubate overnight at 37°C and 5% CO_2.
8. Harvest the cells onto glass fiber filters with the aid of a cell harvester. Determine [^{3}H-methyl] thymidine uptake by liquid scintillation counting.
9. Calculate IL-3 concentration by reference to recombinant IL-3. Using semilog or probit paper, plot standard IL-3 concentration on the x axis vs. proliferation (in cpm) on the y axis. Calculate sample concentration by comparison to standard curve and multiplication by dilution factor.

2. Maintenance of IL-3-Dependent Cell Lines

Materials and Reagents

IL-3-dependent cell line: MC/9 (mouse, ATCC Cat. No. CRL-8306); TF-1 (human, ATCC Cat. No. CRL-2003)

Recombinant IL-3 (mouse or human, depending on origin of indicator cell line) or recombinant GM-CSF (human)

Culture medium: RPMI-1640 containing 10% heat-inactivated FCS and supplemented with 2 mM L-glutamine, 10 mM HEPES, 100 U/ml penicillin, 100 µg/ml streptomycin, 50 µM 2-mercaptoethanol and 1 mM sodium pyruvate

Washing buffer: PBS containing 1% FCS (filter-sterilized)

25-cm^2 culture flasks

15- and 50-ml centrifuge tubes

Pasteur pipettes

Refrigerated centrifuge (set at 4°C)

CO_2 incubator (humidified, set at 37°C and 5% CO_2)

Protocol:

(Note: All materials and reagents must be sterile and proper aseptic technique must be used when handling the cells.)

1. Seed cells at an initial density of 1×10^5 cells/ml in culture medium containing rIL-3 (50 to 100 U/ml). TF-1 cells have been maintained in culture medium containing 1 ng/ml recombinant human GM-CSF.[28]
2. Culture at 37°C and 5% CO_2 for three days. At this point the cells can be harvested and used in the IL-3 assay.
3. Subculture every third day.

Comments:

1. Immune complexes and aggregated immunoglobulins stimulate mast cell lines by binding to their Fc receptors, inducing the production of cytokines (including IL-3) and proliferation.[41]
2. Conditioned medium from the myelomonocytic cell line, WEHI-3 (ATCC Cat. No. TIB 68) can be used as a source of murine IL-3. To prepare, culture WEHI-3 cells at 10^6 cells/ml in culture medium for 48 hours at 37°C and 5% CO_2. Collect supernatant by centrifugation and filter. Aliquot and store at -20°C. Use at a final concentration of 5 to 10%.

3. The human erythroleukemia cell line, TF-1,[40] is responsive to and has been used to assay for IL-5 and GM-CSF. The use of appropriate neutralizing antibodies is required to confirm specificity.

D. INTERLEUKIN-4

Interleukin-4 (IL-4) is a highly pleiotropic cytokine secreted by a subset of CD4[+] T cells (Th2) and mast cells.[42,43] IL-4 has many important activities, including MHC class II molecule and CD23 expression on B cells, the induction of isotype class switch (IgG$_1$, IgE) in B lymphocytes,[44,45] the proliferation and maturation of activated CD4[+] T lymphocytes into the IL-4/IL-5/IL-10-secreting (Th2) subset,[46-48] growth of mast cell lines, activation of some monocytic cell functions, and inhibition of expression of inflammatory cytokines by monocytes.[42,49] IL-4 exerts its effects by interacting with high-affinity IL-4 receptors, which are present on a wide variety of hemopoietic and nonhemopoietic cell types.[50-52] Recently, the functional IL-4R has been shown to be composed of a 140-kDa IL-4-binding subunit[53-55] in noncovalent association with a 70-kDa signal-transducing subunit (IL-2R γ-chain),[25,26] which is shared among other cytokine receptors, including IL-2R,[25,26] IL-7R[27] and possibly IL-13R[28] and IL-15R.[10] In addition to the membrane-bound form, soluble IL-4R forms are also synthesized and released into biological fluids.[53,56]

A wide variety of bioassays have been established for the measurement of IL-4, based on its many activities. Some of these assays include the increased expression of MHC class II molecules on B cells,[57] proliferation of B cells stimulated with anti-immunoglobulin antibodies,[58] secretion of IgG$_1$ or IgE by lipopolysaccharide (LPS)-stimulated B lymphocytes,[59,60] and proliferation of IL-4-responsive T cell and mast cell lines.[9,39] Human and murine IL-4 are species specific in their action. The most commonly used and simple bioassays for murine IL-4 are proliferation assays using IL-4/IL-2-responsive T cell lines, such as HT-2[7] or CTLL-2[8] (described in Section B) or IL-4-responsive, and relatively IL-2-unresponsive T cell lines, such as CT.4S.[61] For the bioassay of human IL-4 murine T cell lines transfected with human IL-4Rs have been reported.[62]

1. CT.4S Proliferation Assay

The HT-2 or CTLL-2 proliferation assay for IL-2 and IL-4 was described in Section B of this chapter. This assay requires the inclusion of neutralizing monoclonal anti-IL-2 and anti-IL-4 to make the assay specific for IL-4 and IL-2, respectively. CT.4S cells, in contrast, exhibit high responsiveness to IL-4 while being unresponsive to IL-2, except at very high concentrations.[61]

This assay, thus, is relatively specific for IL-4 (mouse) and does not require the use of neutralizing monoclonal anti-IL-4 antibodies. Nonetheless, as with all bioassays, it is always a good idea to confirm the specificity of the indicator cell line by inhibition with appropriate neutralizing antibodies.

Materials and Reagents

CT.4S cells (available from Dr. W. E. Paul, NIH)
Recombinant mouse IL-4 standard
Culture medium: RPMI-1640 containing 10% heat-inactivated FCS and supplemented with 2 mM L-glutamine, 10 mM HEPES, 100 U/ml penicillin, 100 µg/ml streptomycin, 50 µM 2-mercaptoethanol and 1 mM sodium pyruvate
Washing buffer: PBS containing 1% FCS (filter-sterilized)
[^{3}H-methyl] thymidine
96-well tissue culture plates (flat bottom)
15- and 50-ml centrifuge tubes
Pasteur pipettes
Refrigerated centrifuge (set at 4°C)
CO_2 incubator (humidified, set at 37°C and 5% CO_2)
Cell harvester (e.g., Skatron)
Glass fiber filter mats (Skatron)
Scintillation vials
Scintillation cocktail (e.g., Econo-Safe™, Research Products International)
Scintillation counter

Protocol:

(Note: All materials and reagents must be sterile and proper aseptic technique must be used when handling the cells.)

1. Dilute test samples and standards (10 to 10,000 pg/ml) in complete medium. Include a negative control containing medium only. Plate at 20 to 50 µl per well (in triplicate).
2. Add enough culture medium to bring the volume of each well to 80 µl.
3. Harvest the CT.4S cells, wash twice with washing buffer and once with culture medium to remove any IL-4. Count and resuspend the cells to a density of 2.5 x 10^5 cells/ml in culture medium. It is advisable to harvest and prepare the indicator cell suspension just before its use.
4. Add 20 µl of the indicator cell suspension to each well (final: 5,000 cells/well).
5. Incubate the plate for 24 hours at 37°C and 5% CO_2.

6. Add 20 µl/well of a 50 µCi/ml solution of [^{3}H-methyl] thymidine in culture medium (final concentration is 1 µCi/well). Incubate overnight at 37°C and 5% CO_2.
7. Harvest the cells onto glass fiber filters with the aid of a cell harvester. Determine [^{3}H-methyl] thymidine uptake by liquid scintillation counting.
8. Calculate IL-4 concentration by comparison to rIL-4 standard curve. Using semilog or probit paper, plot standard IL-4 concentration on the x axis vs. proliferation (in cpm) on the y axis. Calculate sample concentration by comparison to standard curve and multiplication by dilution factor.

2. Maintenance of CT.4S Cells

Materials and Reagents

CT.4S cells
Recombinant IL-4 (mouse)
Culture medium: RPMI-1640 containing 10% heat-inactivated FCS and supplemented with 2 mM L-glutamine, 10 mM HEPES, 100 U/ml penicillin, 100 µg/ml streptomycin, 50 µM 2-mercaptoethanol and 1 mM sodium pyruvate
25-cm^2 culture flasks
15- and 50-ml centrifuge tubes
Pasteur pipettes
Refrigerated centrifuge (set at 4°C)
CO_2 incubator (humidified, set at 37°C and 5% CO_2)

Protocol:

(Note: All materials and reagents must be sterile and proper aseptic technique must be used when handling the cells.)

1. Seed cells at an initial density of 1 x 10^5 cells/ml in culture medium containing rIL-4 (5 to 10 ng/ml).
2. Culture at 37°C and 5% CO_2 for three days. At this point the cells can be harvested by gently shaking the culture flask (cells are adherent) and used in the IL-4 assay.
3. Subculture every third day.

Comments:

1. Murine T cell lines transfected with the human IL-4R (e.g., CTLL)[62] can be used for the bioassay of human IL-4 following a similar protocol, with the exception of human rIL-4 standards. Other bioassays for human IL-4 include the proliferation of human B cells stimulated with anti-IgM antibodies or *Staphylococcus aureus* Cowan strain I (SAC)[63] or class II MHC molecule expression of the human Burkitt lymphoma cell line, RAMOS (ATCC Cat. No. CRL 1596).[64]

E. INTERLEUKIN-5

Interleukin-5 (IL-5) is a cytokine produced primarily by the Th2 subset of CD4+ T cells.[4] Both, in mice and humans, IL-5 appears to be the major cytokine responsible for eosinophilia, promoting the growth, differentiation and activation of eosinophils.[65,66] In addition, murine IL-5 induces proliferation and differentiation of activated B lymphocytes, promotes immunoglobulin secretion, including IgA, and induces differentiation of thymocytes into CTLs.[67-69] Whether human IL-5 has similar effects on human B cells remains controversial. High-affinity receptors for IL-5 are heterodimers composed of an IL-5-binding α-subunit (60 kDa) and a signal-transducing β-subunit (130 kDa) shared with receptors for IL-3 and GM-CSF.[35]

A number of bioassays have been used for the measurement of IL-5, including co-stimulation with dextran sulfate on normal B lymphocytes,[67] proliferation of BCL₁ tumor cells,[70] and induction of IgM secretion by an *in vitro*-adapted clone of BCL₁ cells (BCL₁ CW13.20-3B3).[68] Although BCL₁ cells are also reported to be sensitive to human IL-5, a much simpler proliferation assay using human erythroleukemia TF-1 cells (described in Section C)[40] can also be used for the measurement of human IL-5.

1. Dextran Sulfate Co-Stimulation Assay

Materials and Reagents

Splenic cell suspension from BALB/c mice
Dextran sulfate (Sigma Cat. No. D-7037)
Recombinant mouse IL-5 standard
Culture medium: RPMI-1640 containing 10% heat-inactivated FCS and
    supplemented with 2 mM L-glutamine, 10 mM HEPES, 100 U/ml
    penicillin, 100 µg/ml streptomycin, 50 µM 2-mercaptoethanol and 1 mM
    sodium pyruvate

Washing buffer: PBS containing 1% FCS (filter-sterilized)
[^{3}H-methyl] thymidine
96-well tissue culture plates (flat bottom)
15- and 50-ml centrifuge tubes
Pasteur pipettes
Refrigerated centrifuge (set at 4°C)
CO_2 incubator (humidified, set at 37°C and 5% CO_2)
Cell harvester (e.g., Skatron)
Glass fiber filter mats (Skatron)
Scintillation vials
Scintillation cocktail (e.g., Econo-Safe™, Research Products International)
Scintillation counter

Protocol:

(Note: All materials and reagents must be sterile and proper aseptic technique must be used when handling the cells.)

1. Sacrifice a BALB/c mouse by cervical dislocation and aseptically remove the spleen. Prepare a single cell suspension by gently teasing the spleen with a pair of sterile forceps or by forcing it through a wire-mesh screen. Collect the cell suspension avoiding debris and transfer it to a 50-ml centrifuge tube. Wash the cells twice with 40 ml of cold washing buffer and centrifuge at 200 x g for 10 minutes. Count and resuspend in culture medium to a density of 1 x 10^6 cells/ml.

2. Dilute samples and standards (0.1 to 100 U/ml) in culture medium. Add 100 µl per well, including negative controls containing medium only. Set each dilution in triplicate.

3. Add 50 µl per well of a 80-µg/ml solution of dextran sulfate in culture medium.

4. Add 50 µl per well of the spleen cell suspension and incubate for 72 hours at 37°C and 5% CO_2.

5. Add 20 µl/well of a 50-µCi/ml solution of [^{3}H-methyl] thymidine in culture medium (final concentration is 1 µCi/well) during the last 18 hours of incubation.

6. Harvest the cells onto glass fiber filters with the aid of a cell harvester. Determine [^{3}H-methyl] thymidine uptake.

7. Calculate IL-5 concentration by comparison to rIL-5 standard curve. Using semilog or probit paper, plot standard IL-5 concentration on the x axis vs. proliferation (in cpm) on the y axis. Calculate sample concentration by comparison to standard curve and multiplication by dilution factor.

Comments:

1. A proliferation assay with TF-1 cells (ATCC Cat. No. CRL 2003) can be used as a bioassay for the measurement of human IL-5 (See Section C). This cell line, however, responds also to IL-3, IL-4, IL-6 and GM-CSF, so neutralization with appropriate monoclonal antibodies is advised in order to ascertain specificity.

F. INTERLEUKIN-6

Interleukin-6 (IL-6) is a highly pleiotropic cytokine synthesized by activated T (Th2) and B cells and a variety of non-lymphoid cells, including monocyte/macrophages, fibroblasts, endothelial cells, keratinocytes and several tumor cells.[71] IL-6 plays a central role as a mediator of inflammatory and acute phase responses.[2,71] In addition, IL-6 exerts many other activities, such as induction of proliferation and differentiation of activated B cells, induction of polyclonal immunoglobulin secretion, differentiation of CTLs and promotion of growth of hybridomas, plasmacytomas and several other malignant cells, including Kaposi's sarcoma and multiple myeloma.[71,72] The high affinity receptor for IL-6 is a heterodimer composed of an IL-6-binding α-subunit (80 kDa) that interacts with a signal-transducing β-subunit (130 kDa) that is shared among receptors for IL-6, leukemia-inhibitory factor (LIF) and oncostatin-M.[73] Soluble IL-6 receptors have been reported in the urine and serum of humans.[74]

Although IL-6 has many different activities, its ability to act as a "hybridoma growth factor", inducing the proliferation of a number of IL-6-dependent hybridomas, is the basis for a simple and sensitive assay.

1. Hybridoma Growth Factor Assay

A number of IL-6-dependent hybridomas have been reported. Commonly used indicator cells for the assay of IL-6 are the B9[75] and the 7TD1 hybridomas.[76] Although of murine origin, these indicator cell lines respond to both human and mouse IL-6.

Materials and Reagents

B9 or 7TD1 (ATCC Cat. No. CRL 1851) hybridomas
Recombinant mouse or human IL-6 standard
Culture medium: RPMI-1640 containing 10% heat-inactivated FCS, 2 mM L-glutamine, 10 mM HEPES, 100 U/ml penicillin, 100 μg/ml streptomycin, 50 μM 2-mercaptoethanol and 1 mM sodium pyruvate

97

Washing buffer: PBS containing 1% FCS (filter-sterilized)
[^{3}H-methyl] thymidine
96-well tissue culture plates (flat bottom)
15- and 50-ml centrifuge tubes
Pasteur pipettes
Refrigerated centrifuge (set at 4°C)
CO_2 incubator (humidified, set at 37°C and 5% CO_2)
Cell harvester (e.g., Skatron)
Glass fiber filter mats (Skatron)
Scintillation vials
Scintillation cocktail (e.g., Econo-Safe™, Research Products International)
Scintillation counter

Protocol:

(Note: All materials and reagents must be sterile and proper aseptic technique must be used when handling the cells.)

1. Harvest the indicator cells, wash twice with washing buffer and once with culture medium to remove any IL-6. Centrifuge at 200 x g for 10 minutes. Count and resuspend the cells to a density of 2 x 10⁴ cells/ml in culture medium.
2. Dilute test samples and standards (0.1 to 100 U/ml) in culture medium. Include a negative control containing medium only. Add 100 µl per well (in triplicate).
3. Add 100 µl of the indicator cell suspension to each well (final: 2,000 cells/well).
4. Incubate the plate for 72 hours at 37°C and 5% CO_2.
5. Add 20 µl/well of a 50-µCi/ml solution of [^{3}H-methyl] thymidine in culture medium (final concentration is 1 µCi/well) during the last 4 hours of incubation.
6. Harvest the cells onto glass fiber mats with the aid of a cell harvester. Determine [^{3}H-methyl] thymidine uptake by liquid scintillation counting.
7. Calculate IL-6 concentration by comparison to rIL-6 standard curve. Using semilog or probit paper, plot standard IL-6 concentration on the x axis vs. proliferation (in cpm) on the y axis. Calculate sample concentration by comparison to standard curve and multiplication by dilution factor.

2. Maintenance of B9 or 7TD1 Cells

Materials and Reagents

B9 or 7TD1 (ATCC Cat. No. CRL 1851) hybridomas
Recombinant mouse or human IL-6 standard
Culture medium: RPMI-1640 containing 10% heat-inactivated FCS and
    supplemented with 2 mM L-glutamine, 10 mM HEPES, 100 U/ml
    penicillin, 100 µg/ml streptomycin, 50 µM 2-mercaptoethanol and 1 mM
    sodium pyruvate
25-cm^2 culture flasks
15- and 50-ml centrifuge tubes
Pasteur pipettes
Refrigerated centrifuge (set at 4°C)
CO_2 incubator (humidified, set at 37°C and 5% CO_2)

Protocol:

(Note: All materials and reagents must be sterile and proper aseptic technique
must be used when handling the cells.)

1. Seed cells at an initial density of 5 x 10^4 cells/ml in culture medium
    containing rIL-6 (50 to 100 U/ml).
2. Culture at 37°C and 5% CO_2 until cell density reaches 1 x 10^6 cells/ml
    (approximately 3 to 4 days). At this point the cells should be
    subcultured.
3. For the proliferation assay use cells in the log phase of growth (2 to 3
    days).

Comments:

1. B9 cells have been reported to respond to IL-4, albeit with much lower
    sensitivity compared to IL-6 (approx. 1,000-fold). Potential interference
    with IL-4 can thus be avoided by performing the assay in the presence
    of anti-IL-4 antibodies (11B11, 10 µg/ml). B9 cells are available from
    Dr. Lucien Aarden, Central Laboratory of The Netherlands Red Cross
    Transfusion Service, Amsterdam, The Netherlands).

G. INTERLEUKIN-7

Interleukin-7 (IL-7) is a cytokine produced by bone marrow and thymic stromal cells.[77,78] IL-7 was originally described as a factor that induced proliferation and supported *in vitro* growth of B cell precursors.[77] In addition, IL-7 has been shown to be mitogenic for thymocytes and to co-stimulate with mitogens the proliferation of mature T cells.[79,80] IL-7 functions as a T cell growth factor in ConA blasts or T cell lines the absence of IL-2 or IL-4.[79,80] The effects of IL-7 are mediated by binding to high-affinity IL-7 receptors, which are composed of an α-subunit, responsible for ligand binding, and a β-subunit, responsible for signal transduction.[27,81] The β-subunit is the IL-2R γ chain, which is shared among receptors for IL-2, IL-4, IL-7, IL-13 and IL-15.[10,26-28] Human and murine IL-7 do not exhibit species specificity and are active in either system.[82]

Initially, IL-7 was assayed using progenitor B cells derived from long-term Whitlock-Witte cultures;[83] however, this was both a labor-intensive and a time consuming assay. IL-7 has also been assayed based on its ability to costimulate the proliferation of murine thymocytes in the presence of ConA,[80] however, this assay is not specific for IL-7, as a number of other cytokines, such as IL-1, IL-2, IL-4 and IL-6 also stimulate thymocyte proliferation under these conditions. More recently, IL-7-dependent immature B lymphocytic cell lines, such as 2E8[84] and Ixn/2bx[82] have become available, making possible the assay of IL-7 through a simple proliferation assay.

1. 2E8 Proliferation Assay

This assay is based on the ability of IL-7 to stimulate the proliferation of IL-7-dependent immature B lymphocyte cell line, 2E8.[84] These cells respond to both human and murine IL-7.

Materials and Methods

2E8 cells (ATCC Cat. No. TIB 239)
Recombinant mouse or human IL-7 standard
Culture medium: Iscove's modified DMEM plus 20% FCS, 50 µM 2-mercaptoethanol, 100 U/ml penicillin, 100 µg/ml streptomycin
Washing buffer (PBS-1% FCS)
[^{3}H-methyl] thymidine
96-well tissue culture plates (flat bottom)
15- and 50-ml centrifuge tubes
Pasteur pipettes
Refrigerated centrifuge (set at 4°C)

CO_2 incubator (humidified, set at 37°C and 5% CO_2)
Cell harvester (e.g., Skatron)
Glass fiber filter mats (Skatron)
Scintillation vials
Scintillation cocktail (e.g., Econo-Safe™, Research Products International)
Scintillation counter

Protocol:

(Note: All materials and reagents must be sterile and proper aseptic technique must be used when handling the cells.)

1. Harvest the indicator cells, wash twice with PBS-1% FCS and once with culture medium. Count and resuspend the cells to a density of 1 x 10^5 cells/ml in culture medium.
2. Dilute test samples and standards (0.1 to 100 U/ml) in culture medium. Include a negative control containing medium only. Add 50 µl per well (in triplicate).
3. Add 50 µl of the indicator cell suspension to each well (final: 5,000 cells/well).
4. Incubate the plate for 48 hours at 37°C and 5% CO_2.
5. Add 20 µl/well of a 50-µCi/ml solution of [^{3}H-methyl] thymidine in culture medium (final concentration is 1 µCi/well) during the last 4 hours of incubation.
6. Harvest the cells onto glass fiber mats with the aid of a cell harvester. Determine [^{3}H-methyl] thymidine uptake by liquid scintillation counting.
7. Calculate IL-7 concentration by comparison to rIL-7 standard curve. Using semilog or probit paper, plot standard IL-7 concentration on the x axis vs. proliferation (in cpm) on the y axis. Calculate sample concentration by comparison to standard curve and multiplication by dilution factor.

2. Maintenance of 2E8 Cells

Materials and Reagents

2E8 cells (ATCC Cat. No. TIB 239)
Recombinant mouse or human IL-7 standard
Culture medium: Iscove's modified DMEM plus 20% FCS, 50 µM 2-
    mercaptoethanol, 100 U/ml penicillin, 100 µg/ml streptomycin
Recombinant IL-7 (mouse or human)

25-cm^2 tissue culture flasks
15- and 50-ml centrifuge tubes
Pasteur pipettes
Refrigerated centrifuge (set at 4°C)
CO_2 incubator (humidified, set at 37°C and 5% CO_2)

Protocol:

(Note: All materials and reagents must be sterile and proper aseptic technique must be used when handling the cells.)

1. Seed cells at an initial density of 1 x 10^4 cells/ml in culture medium containing rIL-7 (~1 ng/ml).
2. Culture at 37°C and 5% CO_2 until cell density reaches 5 x 10^5 cells/ml. At this point the cells should be subcultured.
3. For the proliferation assay use cells in the log phase of growth (2 to 3 days).

H. INTERLEUKIN-10

Interleukin-10 (IL-10) is a cytokine produced by a subset of T cells (Th2), certain B cells, EBV-transformed B cell lines and activated monocytes.[85] Originally, IL-10 was described as a product of Th2 cells that inhibited the cytokine secretion by Th1 cells, and was thus termed cytokine-synthesis inhibitory factor (CSIF).[86] However, many other activities have been found to be induced by IL-10, including enhanced proliferation of thymocytes in the presence of IL-2 and IL-4, increased proliferation of mast cells in the presence of IL-3 and IL-4, inhibition of production of inflammatory cytokines (TNFα, IL-1) and antigen-presentation function in macrophages, and induction of growth and differentiation in activated human B cells.[85] An Epstein-Barr virus (EBV)-encoded protein (BCRF1) possesses extensive similarity with IL-10 and exhibits some, but not all, functions associated with IL-10, including cytokine synthesis inhibitory activity.[87]

The original assay for IL-10 was based on its ability to inhibit interferon-γ production by Th1 cell clones or mitogen-stimulated spleen cells.[86] More recently, a simple proliferation assay has been described based on the ability of IL-10 to enhance the proliferation of a mast cell line (MC/9)[39] in the presence of IL-4.[88] Both of these assays are described below. The indicator cells, murine Th1 clones or MC/9 cells, are responsive to both human and mouse IL-10.

1. Cytokine Synthesis Inhibition Assay

In this assay, Th1 cells are first stimulated by antigen plus antigen-presenting cells in the presence of IL-10 standards or samples, followed by measurement of the amount of IFNγ contained in the supernatant after a 24-hour incubation. The presence of IL-10 causes a dose-dependent decrease in IFNγ secretion.

Materials and Reagents

Th1 cell clone (e.g., LB2-1, ATCC Cat. No. CRL 8629)
Recombinant IL-10 standard (human or mouse)
Mitomycin C-treated spleen cell suspension (from a strain syngeneic with Th1 cells; C57Bl/6 for LB2-1)
Antigen (chicken red blood cells for LB2-1)
Recombinant IL-2 (human or mouse)
Culture medium: RPMI-1640 containing 10% heat-inactivated FCS and supplemented with 2 mM L-glutamine, 10 mM HEPES, 100 U/ml penicillin, 100 µg/ml streptomycin, 50 µM 2-mercaptoethanol and 1 mM sodium pyruvate
Washing buffer: PBS containing 1% FCS (filter-sterilized)
96-well tissue culture plates (flat bottom)
15- and 50-ml centrifuge tubes
Pasteur pipettes
Refrigerated centrifuge (set at 4°C)
CO_2 incubator (humidified, set at 37°C and 5% CO_2)
Additional materials for assay of IFNγ (see Section N)

Protocol:

(Note: All materials and reagents must be sterile and proper aseptic technique must be used when handling the cells.)

1.   Prepare a splenic cell suspension, wash once with washing buffer and resuspend in culture medium to a cell density of 1×10^7 cells/ml. Incubate the cells with mitomycin C at a final concentration of 50 µg/ml for 30 minutes at 37°C. Wash the cells four times with 20 ml of ice-cold washing buffer, centrifuging at 200 x g for 10 minutes. Assess viability and resuspend the cells in culture medium to 2×10^7 cells/ml. Alternatively, cells can be irradiated at 2,500 rad.

2.   Harvest "rested" Th1 cells (Th1 cells should be used no earlier than 7 days after the last stimulation). Wash once with washing buffer and

once with culture medium. Count and resuspend to a density of 2×10^6 cells/ml in culture medium containing rIL-2 (4 ng/ml) and the appropriate concentration of antigen (0.04% chicken red blood cells for LB2-1 cells).

3. Dilute samples and standards in culture medium. Add 50 µl per well. Assay each dilution in triplicate.

4. Mix equal volumes of the spleen cell and Th1 cell suspensions (steps # 1 and 2) and add 50 µl per well. Incubate for 24 hours at 37^0C and 5% CO_2.

5. Collect ~75 µl of supernatant from each well, making sure not to get any cells from the bottom.

6. Assay supernatants for IFNγ (see Section N in this chapter).

7. Calculate IL-10 concentration by reference to a rIL-10 standard curve. Using semilog paper, plot standard IL-10 concentration on the x axis vs. IFNγ production on the y axis. Calculate sample concentration by comparison to standard curve and multiplication by dilution factor.

Comments:

1. For a basic protocol for the maintenance of the T cell clones refer to Section A.3 in this chapter. Alternatively, this assay can be performed with ConA-stimulated splenic cells instead of Th1 clones.

2. The presence of IL-4 in the samples may affect this assay due to an increase of IFNγ production, whereas the presence of transforming growth factor-β (TGF-β) may decrease it. If samples are suspected to contain these cytokines, perform the assay in the presence of neutralizing antibodies against these cytokines.

2. MC/9 Proliferation Assay

This assay is based on the ability of IL-10 to enhance the proliferation of MC/9 cells, a mast cell line, in the presence of rIL-4.

Materials and Reagents

MC/9 cells (ATCC Cat. No. CRL 8306)
Recombinant mouse or human IL-10 standard
Recombinant mouse IL-4
Culture medium: RPMI-1640 containing 10% heat-inactivated FCS and supplemented with 2 mM L-glutamine, 10 mM HEPES, 100 U/ml penicillin, 100 µg/ml streptomycin, 50 µM 2-mercaptoethanol and 1 mM sodium pyruvate

Washing buffer (PBS-1% FCS)
[^{3}H-methyl] thymidine
96-well tissue culture plates (flat bottom)
15- and 50-ml centrifuge tubes
Pasteur pipettes
Refrigerated centrifuge (set at 4°C)
CO_2 incubator (humidified, set at 37°C and 5% CO_2)
Cell harvester (e.g., Skatron)
Glass fiber filter mats (Skatron)
Scintillation vials
Scintillation cocktail (e.g., Econo-Safe™, Research Products International)
Scintillation counter

Protocol:

(Note: All materials and reagents must be sterile and proper aseptic technique must be used when handling the cells.)

1. Harvest the indicator cells, wash twice with washing buffer and once with culture medium. Count and resuspend the cells to a density of 5 x 10^4 cells/ml in culture medium containing 400 pg/ml of rIL-4. Background proliferation can be reduced by harvesting the MC/9 cells on the day before the assay, and culturing them overnight (after washing) in culture medium without rIL-3 or the ConA-spleen cell conditioned medium (see protocol for maintenance of MC/9 cells below).
2. Dilute test samples and standards (0.1 to 100 U/ml) in complete medium. Include a negative control containing medium only. Add 100 µl per well (in triplicate).
3. Add 100 µl of the indicator cell suspension to each well (final: 5,000 cells/well).
4. Incubate the plate for 72 hours at 37°C and 5% CO_2.
5. Add 20 µl/well of a 50-µCi/ml solution of [^{3}H-methyl] thymidine in culture medium (final concentration is 1 µCi/well) during the last 4 hours of incubation.
6. Harvest the cells onto glass fiber filters with the aid of a cell harvester. Determine [^{3}H-methyl] thymidine uptake by liquid scintillation counting.
7. Calculate IL-10 concentration by comparison to rIL-10 standard curve. Using semilog or probit paper, plot standard IL-10 concentration on the x axis vs. proliferation (in cpm) on the y axis. Calculate sample

concentration by comparison to standard curve and multiplication by dilution factor.

3. Maintenance of MC/9 Cells

Materials and Reagents

MC/9 cells (ATCC Cat. No. CRL 8306)
Recombinant mouse IL-3 or ConA-activated splenic cell conditioned medium
Culture medium: RPMI-1640 containing 10% heat-inactivated FCS and supplemented with 2 mM L-glutamine, 10 mM HEPES, 100 U/ml penicillin, 100 µg/ml streptomycin, 50 µM 2-mercaptoethanol and 1 mM sodium pyruvate
25-cm^2 tissue culture flasks
15- and 50-ml centrifuge tubes
Pasteur pipettes
Refrigerated centrifuge (set at 4°C)
CO_2 incubator (humidified, set at 37°C and 5% CO_2)

Protocol:

(Note: All materials and reagents must be sterile and proper aseptic technique must be used when handling the cells.)

1. Seed cells at an initial density of 5×10^4 cells/ml in culture medium containing 5% ConA-activated splenic cell-conditioned medium (a source of growth factors). To prepare the ConA-supernatant, culture murine spleen cells at a final cell density of 1×10^7 cells/ml in culture medium containing ConA (5 µg/ml) for 24 hours at 37°C and 5% CO_2. Harvest by centrifugation, filter and store at -20°C.
2. Culture at 37°C and 5% CO_2 for 3 days. At this point the cells should be subcultured.
3. For the proliferation assay use cells in the log phase of growth (2 to 3 days).

Comments:

1. IL-3 and IL-4 stimulate proliferation of mast cells and synergize with IL-10. Their presence in samples may affect the results. MC/9 cells also respond to murine and human stem cell factor (SCF or c-*kit* ligand).

106

I. INTERLEUKIN-12

Interleukin-12 (IL-12) is a heterodimeric cytokine produced by a variety of cells, including macrophages, B cells and keratinocytes, that was originally described by its ability to stimulate cytotoxicity of NK cells and maturation of CTLs.[89,90] IL-12 has multiple effects on T cells and NK cells, inducing the secretion of IFNγ from resting and activated cells and acting as a co-stimulant for the proliferation of T cells in the presence of mitogens.[89,91] Partly because of its ability to induce IFNγ synthesis, IL-12 has been shown to promote the development of Th1 responses and to inhibit that of Th2.[92]

Bioassays for IL-12 were originally based on its ability to stimulate the production of IFNγ from unstimulated peripheral blood mononuclear cells (PBMC), or to stimulate the proliferation of human PHA-stimulated T cell blasts.[89,91] This latter assay, however, lacks specificity for IL-12, making necessary the inclusion of neutralizing anti-IL-12 antibodies (when assaying samples potentially containing other cytokines in addition to IL-12) in order to confirm that the observed proliferation is due to IL-12. A modification of the PHA-blast proliferation assay has been reported,[93] in which samples are first incubated in wells containing immobilized anti-IL-12 antibodies, and then the captured IL-12 is detected by adding PHA-blasts and measuring proliferation.

1. IFNγ Induction Assay for IL-12

In this assay, human PBMC are cultured in the presence of IL-12-containing samples, followed by collection of the supernatants which are later assayed for IFNγ.

Materials and Reagents

Suspension of human peripheral blood mononuclear cells (PBMC)
Histopaque-1077 (Sigma Cat. No. H-8889)
Blood collection tubes (heparinized)
Recombinant (human) IL-12 standard
Culture medium: RPMI 1640 containing 10% heat-inactivated FCS and
    supplemented with 2 mM L-glutamine, 10 mM HEPES, 100 U/ml
    penicillin, 100 µg/ml streptomycin, 50 µM 2-mercaptoethanol and 1 mM
    sodium pyruvate
Washing buffer: PBS containing 1% FCS (filter-sterilized)
Sterile PBS
100 x 15-mm Petri dishes
96-well tissue culture plates (flat bottom)

15- and 50-ml centrifuge tubes
Pasteur pipettes
Refrigerated centrifuge (set at 4°C)
CO_2 incubator (humidified, set at 37°C and 5% CO_2)
Additional materials for assay of IFNγ (see Section N in this chapter)

Protocol:

(Note: All materials and reagents must be sterile and proper aseptic technique must be used when handling the cells.)

1. Collect blood (heparinized) from healthy human donors and isolate PBMC by density gradient centrifugation using Histopaque-1077 (see Chapter 1). Wash cells in sterile PBS, count and resuspend in culture medium at 5 x 10^6 cells/ml and deplete adherent cells by incubation on plastic petri dishes for 1 hour at 37°C. Repeat adherence procedure once more.
2. Resuspend PBMC in culture medium to a density of 4 x 10^6 cells/ml. Add 100 µl per well.
3. Dilute samples and standards (20 to 2,000 pg/ml) with culture medium. Test each dilution in triplicate, including a negative control containing medium only. Add 100 µl per well.
4. Incubate at 37°C and 5% CO_2 for 12 hours.
5. Remove approximately 100 µl of supernatant making sure not to take any cells from the bottom of the plate.
6. Assay supernatants for human IFNγ (see IFNγ assay later in this Chapter).
7. Calculate IL-12 concentration. Plot standard IL-12 concentration on the x axis vs. IFNγ production on the y axis. Calculate sample concentration by comparison to standard curve and multiplication by dilution factor.

2. Lymphoblast Proliferation Assay for IL-12

This assay is based on the ability of IL-12 to induce proliferation of PHA-activated human T cell blasts. As other cytokines, such as IL-2, IL-4 could also score in this assay, neutralization of IL-12 with monoclonal antibodies is recommended in order to confirm that the response is due to IL-12. This assay can detect either human or murine IL-12.

Materials and Reagents

Suspension of PBMC
Histopaque-1077 (Sigma Cat. No. H-8889)
Blood collection tubes (heparinized)
Recombinant (human or mouse) IL-12 standard
Phytohemagglutinin-P (PHA, Sigma Cat. No. L-9132)
Culture medium: RPMI-1640 containing 10% heat-inactivated FCS and
    supplemented with 2 mM L-glutamine, 10 mM HEPES, 100 U/ml
    penicillin, 100 µg/ml streptomycin, 50 µM 2-mercaptoethanol and 1 mM
    sodium pyruvate
Washing buffer: PBS containing 1% FCS (filter-sterilized)
Sterile PBS
[^{3}H-methyl] thymidine
75-cm^2 tissue culture flasks
96-well tissue culture plates (flat bottom)
15- and 50-ml centrifuge tubes
Pasteur pipettetes
Refrigerated centrifuge (set at 4°C)
CO_2 incubator (humidified, set at 37°C and 5% CO_2)
Cell harvester (e.g., Skatron)
Glass fiber filter mats (Skatron)
Scintillation vials
Scintillation cocktail (e.g., Econo-Safe™, Research Products International)
Scintillation counter

Protocol:

(Note: All materials and reagents must be sterile and proper aseptic technique
must be used when handling the cells.)

1. Prepare PBMC by density gradient centrifugation, as in previous
   section. Wash three times with sterile PBS, assess viability and
   resuspend in culture medium to a density of 5 x 10^5 cells/ml.
2. Add enough PHA from a 5 mg/ml stock solution in culture medium to
   the cell suspension, in order to obtain a final PHA concentration of 5
   µg/ml. Culture for 72 hours at 37°C and 5% CO_2 using 75-cm^2 tissue
   culture flasks.
3. Split cells 1:2 with culture medium. Add rIL-2 (human) to a final
   concentration of 50 U/ml. Incubate for an additional 24 hours.
4. Harvest the cells into 50-ml centrifuge tubes, wash twice with ice-cold

washing buffer and once with culture medium in order to remove the rIL-2.

5. Count the cells and resuspend the lymphoblasts in culture medium to a cell density of 4 x 10^5 cells/ml. Add 50 µl per well.

6. Dilute samples and standards (20 to 2,000 pg/ml) with culture medium. Add 50 µl per well. Include negative controls containing medium alone. Assay each dilution in triplicate.

7. Incubate for 24 hours at 37°C and 5% CO_2.

8. Add 20 µl/well of a 50-µCi/ml solution of [^{3}H-methyl] thymidine in culture medium (final concentration is 1 µCi/well) during the last 4 to 6 hours of incubation.

9. Harvest the cells onto glass fiber mats with the aid of a cell harvester. Determine [^{3}H-methyl] thymidine uptake by liquid scintillation counting.

10. Calculate IL-12 concentration by comparison to rIL-12 standard curve. Using semilog or probit paper, plot standard IL-12 concentration on the x axis vs. proliferation (in cpm) on the y axis. Calculate sample concentration by comparison to standard curve and multiplication by dilution factor.

Comments:

1. This assay detects both human and mouse IL-12. Since the indicator cells are of human origin, the possible interference by murine IL-2 or IL-4 (which are not active on human cells) is avoided. This, however, is not the case with human samples.

J. INTERLEUKIN-13

Interleukin-13 (IL-13) is a cytokine produced primarily by subsets of T cells (Th2 cells).[94] IL-13 has moderate sequence homology (25 to 30%) to IL-4, and shares with it a number of biological activities in the human, including upregulation of B cell class II MHC and CD23 molecule expression, B cell proliferation and secretion of IgE, and inhibition of production of inflammatory cytokines by macrophages.[94-96] In the mouse, however, IL-13 appears to modulate only the monocytic lineage.[97] Moreover, the IL-13 receptor shares a common signal-transducing subunit (the IL-2R γ chain) with receptors for IL-2, IL-4, IL-7 and IL-15.[10,26-28]

The bioassays for IL-13 is based on its ability to induce the proliferation of the B9 hybridoma.[98] This is basically the same proliferation assay used for the assay of IL-6 (see Section F in this chapter). B9 cells also respond to IL-

6 and to a lesser extent, IL-4. Both human and mouse IL-13 are detected by this assay.

1. B9 Proliferation Assay for IL-13

Materials and Reagents

B9 hybridoma
Recombinant mouse or human IL-13 standard
Culture medium: RPMI-1640 containing 10% heat-inactivated FCS and supplemented with 2 mM L-glutamine, 10 mM HEPES, 100 U/ml penicillin, 100 µg/ml streptomycin, 50 µM 2-mercaptoethanol and 1 mM sodium pyruvate
Washing buffer: PBS containing 1% FCS (filter-sterilized)
[^{3}H-methyl] thymidine
96-well tissue culture plates (flat bottom)
15- and 50-ml centrifuge tubes
Pasteur pipettes
Refrigerated centrifuge (set at 4°C)
CO_2 incubator (humidified, set at 37°C and 5% CO_2)
Cell harvester (e.g., Skatron)
Glass fiber filter mats (Skatron)
Scintillation vials
Scintillation cocktail (e.g., Econo-Safe™, Research Products International)
Scintillation counter

Protocol:

(Note: All materials and reagents must be sterile and proper aseptic technique must be used when handling the cells.)

1. Harvest the indicator cells, wash twice with washing buffer and once with culture medium to remove any rIL-6. Count and resuspend the cells to a density of 1 x 10^5 cells/ml in culture medium.
2. Dilute test samples and standards (20 to 2,000 pg/ml) in culture medium. Include a negative control containing medium only. Add 50 µl per well (in triplicate).
3. Add 50 µl of the indicator cell suspension to each well (final: 5,000 cells/well).
4. Incubate the plate for 48 to 72 hours at 37°C and 5% CO_2.
5. Add 20 µl/well of a 50-µCi/ml solution of [^{3}H-methyl] thymidine in culture medium (final concentration is 1 µCi/well) during the last 4 to 6 hours of incubation.

6. Harvest the cells onto glass fiber mats with the aid of a cell harvester. Determine [³H-methyl] thymidine uptake by liquid scintillation counting.
7. Calculate IL-13 concentration by comparison to rIL-13 standard curve. Using semilog or probit paper, plot standard IL-13 concentration on the x axis vs. proliferation (in cpm) on the y axis. Calculate sample concentration by comparison to standard curve and multiplication by dilution factor.

Comments:

1. For the maintenance of B9 cells, refer to Section F in this chapter.
2. B9 cells proliferate also in response to IL-6 and IL-4. When assaying murine samples, include neutralizing anti-IL-6 and anti-IL-4 antibodies if the presence of these cytokines is suspected. When assaying human samples, only anti-human IL-6 antibodies are required, as human IL-4 is not active on murine cells.
3. As with other cytokine assays, confirmation of IL-13 activity by inhibition with anti-IL-13 antibodies is recommended.

K. INTERLEUKIN-15

Interleukin-15 (IL-15) is a newly described cytokine produced by a variety of cells, including adherent PBMC, epithelial and fibroblast cell lines.[10] IL-15 shares many activities with IL-2, including the proliferation of activated human and mouse T cells and generation of LAK cells.[10,99] The IL-15 receptor appears to share with the IL-2R, both the β- and γ-subunits, since antibodies directed against the IL-2Rβ subunit block both IL-2 and IL-15 activity.[10]

The bioassay for IL-15 is based on its ability to stimulate the proliferation of CTLL-2 cells.[10] Inasmuch as IL-2 also stimulates the proliferation of this cell line, identification of IL-15, requires its neutralization with anti-IL-15 or anti-IL-2Rβ chain monoclonal antibodies, but not by anti-IL-2 or IL-2Rα antibodies.

1. CTLL-2 Proliferation Assay for IL-15

Materials and Reagents

CTLL-2 cells (ATCC Cat. No. TIB 214)
Recombinant IL-15 standard (simian rIL-15 is available from Genzyme)
Rat monoclonal anti-mouse IL-4 (11B11) (ATCC Cat. No. HB 188)

Rat monoclonal anti-mouse IL-2 (S4B6) (ATCC Cat. No. HB 8794)
Rat monoclonal anti-mouse IL-2Rβ (Pharmingen, Cat. No. 01921D)
Culture medium: RPMI-1640 containing 10% heat-inactivated FCS and supplemented with 2 mM L-glutamine, 10 mM HEPES, 100 U/ml penicillin, 100 µg/ml streptomycin, 50 µM 2-mercaptoethanol and 1 mM sodium pyruvate
Washing buffer: PBS containing 1% FCS (filter-sterilized)
[^{3}H-methyl] thymidine
96-well tissue culture plates (flat bottom)
15- and 50-ml centrifuge tubes
Pasteur pipettes
Refrigerated centrifuge (set at 4°C)
CO_2 incubator (humidified, set at 37°C and 5% CO_2)
Cell harvester (e.g., Skatron)
Glass fiber filter mats (Skatron)
Scintillation vials
Scintillation cocktail (e.g., Econo-Safe™, Research Products International)
Scintillation counter

Protocol:

(Note: All materials and reagents must be sterile and proper aseptic technique must be used when handling the cells.)

1. Harvest the CTLL-2 cells (see Section B, in this chapter), wash twice with cold washing buffer and once with complete medium in order to remove any IL-2. Count and resuspend the cells in culture medium to 2.5 x 10^5 cells/ml.
2. Dilute samples and standards (1 to 100 ng/ml) in culture medium. Assay each dilution in triplicate. Include a negative control containing medium alone.
3. Add 20 to 50 µl of sample or recombinant IL-15 standard per well.
4. In the case of assays involving murine samples, the assay can be made specific for IL-15 by including neutralizing anti-IL-2 and anti-IL-4 antibodies (final concentration 10 µg/ml). In the case of human samples, only anti-human IL-2 antibodies need to be used (human IL-4 is not active in murine systems). Preincubate samples and antibodies for 30 minutes at room temperature.
5. Add enough culture medium to bring the volume of each well to 80 µl.
6. Add 20 µl of the indicator cell suspension to each well (5,000 cells/well). The final volume is 100 µl. Incubate for 24 hours at 37°C in a 5% CO_2 atmosphere.

7. Add 20 µl/well of a 50-µCi/ml solution of [³H-methyl] thymidine in culture medium (final concentration is 1 µCi/well) during the last 4 to 6 hours of incubation.
8. Harvest the cells onto glass fiber mats with the aid of a cell harvester. Determine [³H-methyl] thymidine uptake by liquid scintillation counting.
9. Calculate IL-15 concentration by comparison to rIL-15 standard curve. Using semilog or probit paper, plot standard IL-15 concentration on the x axis vs. proliferation (in cpm) on the y axis. Calculate sample concentration by comparison to standard curve and multiplication by dilution factor.

Comments:

1. Refer to Section B in this chapter for the maintenance of CTLL-2 cells.
2. Proliferation of CTLL-2 cells to IL-15 is blocked by antibodies against the IL-2R β, but not against the IL-2R α-chain.
3. This assay detects both human and mouse IL-15.

L. OTHER CYTOKINES

1. Chemokines, including Interleukin-8

Chemokines are a family of "small cytokines" (8 to 10 kDa) that play a central role in inflammation, by functioning as chemoattractants and activating factors for a variety of cells, including granulocytes, monocytes, and T lymphocytes.[2,100] Based on amino acid sequence similarities, chemokines have been divided into two different subgroups: the α or C-X-C group (their first two cysteines are separated by a single amino acid) and the β or C-C group (their first two cysteine residues are adjacent to each other). The C-X-C group includes cytokines such as IL-8 or NAP-1 (neutrophil attractant protein-1), MGSA/GRO, PF-4, NAP-2, and IP 10. The C-C group includes MCAF, RANTES, LD-78, ACT-2 and I-309.[2,100] IL-8 and other chemokines are produced by many cell types in response to activation by mitogens, infectious agents or other cytokines, particularly TNFα and IL-1.[2,101]

Since many different chemokines exist and they overlap each other in their actions and chemoattractant properties, individual chemokines are best assayed by immunoassays, based on reactivity with specific antibodies. However, IL-8 and other chemokines can also be assayed based on their chemotactic properties for granulocytes or monocytes in multiwell chemotactic chambers.[102]

2. Interleukin-9

Interleukin-9 (IL-9, P40/TGFIII) is a cytokine produced by T lymphocytes.[103,104] IL-9 has growth factor activity on a variety of cells, including T cells, mast cells and myeloid and erythroid precursor cells.[104,105] A bioassay for IL-9 is based on its ability to stimulate the proliferation of an IL-9-dependent megakaryocytic cell line (M-07e).[104] This cell line is dependent on either human GM-CSF or IL-3 for growth. The reagents necessary for the IL-9 bioassay, including the indicator cell line, recombinant IL-9 standard and neutralizing anti-IL-9 antibodies are available to academic laboratories from Genetics Institute (Cambridge, MA). Other commercial sources of rIL-9 are also available. This assay detects both human and mouse IL-9.

3. Interleukin-11

Interleukin-11 (IL-11) is a cytokine originally described as a product of a bone marrow stromal cell line that induced proliferation of an IL-6-dependent plasmacytoma cell line (T1165).[106] In addition to this activity, IL-11 has been demonstrated to have a growth-promoting effect on some hemopoietic precursor cells (including megakaryocytic and erythroid lineages) and lymphoid cells.[106] IL-11 appears to share some functions in common with IL-6, such as acting as an adipogenesis-inhibiting factor and inducing synthesis of acute-phase proteins by liver cells.[107] The IL-11 receptor shares the gp130 signal-transducing subunit with the IL-6R.[108]

A bioassay for IL-11 is based on its ability to stimulate the proliferation of murine T10 cells, a subline derived from the parental T1165 cell line.[109] T10 cells are maintained in the presence of human rIL-11 and respond to both human and mouse IL-11. The reagents for this assay, including the indicator cell line, recombinant IL-11 standard and neutralizing anti-IL-11 antibodies are available to academic laboratories from Genetics Institute (Cambridge, MA). Other commercial sources of rIL-11 are also available.

M. TUMOR NECROSIS FACTORS

Tumor necrosis factor-α (TNFα) and tumor necrosis factor-β (TNFβ, lymphotoxin) are two pleiotropic cytokines secreted primarily by monocyte/macrophages and T lymphocytes, respectively.[110] TNFα was originally described as a factor present in the serum of *Bacillus Calmette-Guerin* (BCG)-treated mice that induced tumor necrosis in tumor bearing mice.[111] Although there is only a moderate degree of homology between TNFα and TNFβ, both forms bind to the same receptors and elicit similar activities.[110,112] TNFα is the principal mediator of natural immunity against Gram-negative bacteria and a key mediator of inflammatory responses and septic shock.[110] In addition, TNFα/β have many other activities, including a cytotoxic effect towards certain target cells and tumors, induction of MHC class I and II molecules on target cells, activation of polymorphonuclear leukocytes, induction of expression of adherence molecules on endothelial cells, and co-stimulatory effects of T and B lymphocytes.[2,110,112] Extracellular forms of TNF receptors are shed and appear in biological fluids, potentially acting as regulators of TNF activity.[113-115]

Bioassays for TNFα/β are based on its cytotoxic effect toward target cells, such as murine L929 fibroblasts[116] and WEHI-164 fibrosarcoma cell line.[117] Although these bioassays detect both human and mouse TNF activity, they do not differentiate between TNFα and TNFβ.

1. L929 Cytotoxicity Assay for TNFα/β

Materials and Reagents

Murine L929 fibroblasts (TNF-sensitive)
Recombinant TNFα or TNFβ standards (human or mouse)
Actinomycin D (Sigma Cat. No. A-9415)
Culture medium: RPMI-1640 containing 10% heat-inactivated FCS and
    supplemented with 2 mM L-glutamine, 10 mM HEPES, 100 U/ml
    penicillin, 100 µg/ml streptomycin, 50 µM 2-mercaptoethanol and 1 mM
    sodium pyruvate
Porcine trypsin solution (Sigma Cat. No. T-4549)
Washing buffer: PBS containing 1% FCS (filter-sterilized)
Reagents necessary for MTT proliferation assay (see Chapter 3)
8-well aspirator (e.g., Drummond, CMS)
96-well tissue culture plates (flat bottom)
15- and 50-ml centrifuge tubes
Pasteur pipettes
Refrigerated centrifuge (set at 4°C)
CO_2 incubator (humidified, set at 37°C and 5% CO_2)
Multiwell spectrophotometer (ELISA reader)

Protocol:

(Note: All materials and reagents must be sterile and proper aseptic technique must be used when handling the cells.)

1. On day one, harvest L929 fibroblasts by trypsinization (see protocol below). Remove trypsin by washing cells once in 20 ml of cold culture medium. Centrifuge the cells at 200 x g for 10 minutes, count and resuspend in culture medium to a cell density of 4 x 10^5 cells/ml.
2. Using a 96-well tissue culture plate, add 100 µl of the fibroblast suspension per well and incubate overnight at 37°C and 5% CO_2. Set one horizontal row for each sample to be tested, plus one for the standards.
3. Next day, check the cells with an inverted microscope. Under these conditions, cells should be confluent. Using an 8-channel aspirator, aspirate the medium from the wells.
4. Dilute samples and standards (serial twofold dilutions) in culture medium. Add 50 µl of each dilution per well, starting with column 2. Add 50 µl of medium alone to column 1 (medium control).
5. Add 50 µl/well of a 2-µg/ml solution of Actinomycin D in culture medium. Incubate the plates for 18 hours at 37°C and 5% CO_2.
6. Add 10 µl per well of a 5 mg/ml solution of MTT in PBS. Return to the incubator and incubate for 4 additional hours.
7. Remove supernatants from the wells with an 8-channel aspirator. Add 100 µl per well of acidic isopropanol (0.04 N HCl in isopropanol). Mix well until all crystals are dissolved.
8. Read on a multiwell spectrophotometer using a test wavelength of 570 nm and a reference wavelength of 630 nm. Read plates within one hour of addition of the acidic isopropanol. (Refer to Chapter 3 for MTT assay.)
9. Using semilog paper, plot OD_{570} (in y axis) versus the standard concentration (in the x axis). Calculate TNF concentration in unknown samples by referring to standard curve and multiplying by the dilution factor. Alternatively, calculation can also be performed by calculating the dilution of sample giving 50% of maximal lysis and comparing that to the standard dilution at which 50% maximal lysis occurs.

Comments:

1. L929 cells from different sources may vary widely in their sensitivity to TNF. It is better to obtain cells whose TNF-sensitivity is known.
2. Several modifications to the original MTT proliferation assay are discussed in Chapter 3. The extraction reagent (acidic isopropanol) can be substituted for 10% SDS (pH 4.5) or other reagents to minimize protein precipitation if it interferes with spectrophotometric readings.

2. Maintenance and Passage of L929 Cells

Materials and Reagents

L929 fibroblasts
Diluted porcine trypsin solution (Sigma Cat. No. T-4549)
Culture medium: RPMI-1640 containing 10% heat-inactivated FCS and supplemented with 2 mM L-glutamine, 10 mM HEPES, 100 U/ml penicillin, 100 µg/ml streptomycin, 50 µM 2-mercaptoethanol and 1 mM sodium pyruvate
75-cm^2 tissue culture flasks
15- and 50-ml centrifuge tubes
Pasteur pipettes
Refrigerated centrifuge (set at 4°C)
CO_2 incubator (humidified, set at 37°C and 5% CO_2)

Protocol:

(Note: All materials and reagents must be sterile and proper aseptic technique must be used when handling the cells.)

1. Seed fibroblasts at an initial density of 4 x 10^5 cells/ml in culture medium.
2. Incubate at 37°C and 5% CO_2 until the cells reach confluence (~5 to 7 days). The medium can be discarded and replaced on day 4.
3. For subculture, aspirate the medium and add 3 ml (for a 75-cm^2 flask) of a trypsin solution. Incubate at 37°C for approximately 1 to 2 minutes or until cells start to detach from the flask.
4. Shake the flask and transfer the cells with a sterile pipettete into a 50-ml centrifuge tube containing 20 ml of culture medium at 4°C. Resuspend in culture medium and count. The cells are ready for subculture or assay.

N. INTERFERONS

Interferons (IFN) are a group of cytokines with the ability to inhibit viral replication *in vitro*.[2] Interferons have been divided into two main groups: Type I or viral interferon, including IFNα (leukocyte derived) and IFNβ (fibroblast derived); and Type II or immune interferon, including IFNγ, which is produced mainly by T lymphocytes as a result of antigenic or mitogenic stimulation.[118] IFNα and IFNβ have similar activities and bind to the same membrane receptors,[119] whereas IFNγ, which displays a different set of immunoregulatory activities, binds to a different receptor.[120] The activities of IFNγ are many, and include induction of expression of MHC class I and II molecules on a variety of target cells, macrophage activation, increased cytotoxicity in NK cells and CTLs, inhibition of many IL-4-mediated functions and effects on B cell differentiation.[121]

Interferons, in general, can be assayed based on their antiviral activity.[122] This assay, however, does not distinguish among the different types of IFN, although with the use of specific neutralizing antibodies, this may be possible. In addition, several bioassays, not based on antiviral activity, have been reported for the detection and quantitation of IFNγ. Some of these assays include the induction of class II MHC molecule expression on target cells,[123] induction of microbicidal activity,[124] and cytotoxicity against target cells such as the murine B lymphoma cell line, WEHI-279.[125]

1. Antiviral Activity Assay for Interferons α, β and γ

The basis for this assay is the antiviral effect induced by all interferons. In this assay, murine L929 fibroblasts are cultured in the presence of vesicular stomatitis virus (VSV) and dilutions of IFN-containing samples and standards. Measurement of IFN activity is then based on the inhibition of the cytopathic effect (cell lysis). This assay employs the MTT reduction assay to quantitate the extent of cell death.

Materials and Reagents

Murine L929 fibroblasts (IFN-sensitive)
Vesicular stomatitis virus (VSV, Indiana strain; ATCC)
Recombinant IFNα, β or γ standard (mouse)
Culture medium: Eagle minimum essential medium, EMEM, containing 5% FCS, 15 mM HEPES, 100 U/ml penicillin, 100 µg/ml streptomycin
Trypsin solution (Sigma Cat. No. T-4549)
Reagents needed for MTT assay (see Chapter 3)
8-well aspirator (e.g., Drummond, CMS)
96-well tissue culture plates
75-cm^2 tissue culture flasks
15- and 50-ml centrifuge tubes
Pasteur pipettes
Refrigerated centrifuge (set at 4°C)
CO_2 incubator (humidified, set at 37°C and 5% CO_2)

119

Protocol:

(Note: All materials and reagents must be sterile and proper aseptic technique must be used when handling the cells.)

1. Harvest L929 cells after they have reached confluence (see protocol for maintenance of L929 cells, Section M in this chapter). Seed enough cells into a new flask so that they will again be confluent in 24 hours (~1:2 dilution in culture medium). Use these "new" cells for the assay.
2. Harvest the cells by trypsinization. Wash three times in culture medium in order to remove the trypsin. Centrifuge at 200 x g for 10 minutes, count and resuspend to a cell density of 1.5×10^5 cells/ml in culture medium.
3. Add 100 μl of medium to each well. Allow one row for each sample and one for the standards.
4. Add 50 μl of sample to the third well in each row. Prepare serial threefold dilutions by mixing and transferring 50 μl to the next well (well # 4) and repeating until well # 12. Discard the last 50 μl. The first two wells are reserved for controls: cells alone (well # 1) and cells plus virus without sample (well # 2).
5. Add 100 μl of the fibroblast suspension to each well, except for a well reserved as a blank.
6. Incubate the cells for 24 hours at 37°C and 5% CO_2.
7. On the next day, dilute the VSV suspension in culture medium to 10^3 PFU/ml.
8. Aspirate medium from the wells using an 8-channel aspirator. Add 200 μl of the virus suspension to all wells except for the first column (cells only). Add 200 μl of culture medium to these wells instead.
9. Incubate the plates at 37°C and 5% CO_2 for at least 24 hours. Observe the cells under an inverted microscope to determine the amount of cell destruction. Cells cultured with VSV in the absence of IFN should be 75 to 90% destroyed.
10. Using an 8-channel aspirator, aspirate the supernatant from the plates into a flask containing bleach. Add 80 μl per well of culture medium and 20 μl per well of a 5-mg/ml MTT solution in PBS, except for the blank.
11. Incubate for 4 hours at 37°C and 5% CO_2, then add 100 μl per well of an acidic isopropanol solution. Mix and allow crystals to dissolve. (Refer to Chapter 3 for MTT proliferation assay.)
12. Read on a multiwell scanning spectrophotometer using a test wavelength of 570 nm and a reference wavelength of 630 nm. Read plates within one hour of addition of the acidic isopropanol. Use the

blank for zeroing the spectrophotometer.

13. Calculate IFN concentration. Plot OD_{570} (y axis) as a function of standard IFN concentration (x axis). Calculate IFN concentration in unknown samples by referring to standard curve and multiplying by the dilution factor. Alternatively, calculation can also be performed by calculating the dilution of sample giving 50% of maximal lysis and comparing that to the standard dilution at which 50% maximal lysis occurs.

Comments:

1. As with TNFα, the sensitivity of L929 cells from different sources to IFN varies widely. Obtain cells of known sensitivity.
2. The appropriate dose of virus is crucial to the sensitivity and reproducibility of the assay. An appropriate dose should give detachment of 75 to 90% of cells after 24 hours. It is advisable to titrate virus pools to determine the optimal challenge dose.
3. VSV is a potentially pathogenic virus. Appropriate care and technique should be used when handling the virus or VSV-containing cultures.
4. As interferons are species specific, this assay detects murine but not human IFN. The bioassay for human IFN can be performed almost identically with the exception of using susceptible human cell lines and recombinant human IFN standards. Some human cells lines used with VSV are: WISH, a human amnion epithelial cell line (ATCC Cat. No. CCL 25), AG-1732 and GM-2504A, a pair of human skin fibroblast cell lines.[126]
5. Stocks of VSV are generated from the supernatants of infected L929 or Vero cells. Supernatants are aliquoted and stored frozen at -70°C. The potency of the virus stock is determined by a plaque assay method.[126]

2. WEHI-279 Cell Cytotoxicity Assay for Murine IFNγ

This assay is based on the ability of IFNγ to inhibit the proliferation of the murine B lymphoma cell line, WEHI-279.[125] The indicator cells are cultured in the presence of IFNγ-containing samples or standards, and the inhibition of proliferation is assessed by [³H-methyl] thymidine incorporation.

Materials and Reagents

WEHI-279 cells (ATCC Cat. No. CRL-1704)
Recombinant murine IFNγ standard
Culture medium: RPMI-1640 containing 10% heat-inactivated FCS and

supplemented with 2 mM L-glutamine, 10 mM HEPES, 100 U/ml penicillin, 100 µg/ml streptomycin, 50 µM 2-mercaptoethanol and 1 mM sodium pyruvate
Washing buffer: PBS containing 1% FCS (filter-sterilized)
[^{3}H-methyl] thymidine
96-well tissue culture plates (flat bottom)
15- and 50-ml centrifuge tubes
Pasteur pipettes
Refrigerated centrifuge (set at 4°C)
CO_2 incubator (humidified, set at 37°C and 5% CO_2)
Cell harvester (e.g., Skatron)
Glass fiber filter mats (Skatron)
Scintillation vials
Scintillation cocktail (e.g., Econo-Safe™, Research Products International)
Scintillation counter

Protocol:

(Note: All materials and reagents must be sterile and proper aseptic technique must be used when handling the cells.)

1. Harvest the WEHI-279 cells. Wash twice with 20 ml of cold washing buffer, centrifuging at 200 x g for 10 minutes. Count the cells and resuspend in culture medium to a density of 4 x 10^5 cells/ml.
2. Dilute samples and standards (0.1 to 100 U/ml) in culture medium. Add 50 µl/well. Include a control containing medium alone. Assay each dilution in triplicate.
3. Add 50 µl/well of the indicator cell suspension. Incubate for 24 hours at 37°C and 5% CO_2.
4. Add 20 µl/well of a 50-µCi/ml solution of [^{3}H-methyl] thymidine in culture medium (final concentration is 1 µCi/well) during the last 4 to 6 hours of incubation.
5. Harvest the cells onto glass fiber mats with the aid of a cell harvester. Determine [^{3}H-methyl] thymidine uptake by liquid scintillation counting.
6. Calculate IFNγ concentration by comparison to IFNγ. standard curve. Using semilog or probit paper, plot standard IFNγ concentration on the x axis vs. proliferation (in cpm) on the y axis. Calculate sample concentration by comparison to standard curve and multiplication by dilution factor.

Comments:

1. WEHI-279 cells are maintained by culture in RPMI-10% FCS medium. Seed flask at an initial density of 5 x 10⁴ cells/ml. Subculture every third day or when the cells reach an approximate density of 1 x 10⁶ cells/ml. Culturing cells at higher densities will result in decreased cell viability and potential changes in their sensitivity to IFNγ.

2. The sensitivity of WEHI-279 cells to IFNγ may decrease after prolonged culture. It is advisable to keep aliquots of frozen cells so fresh cultures can be started if such is the case.

3. As with other cytokines, results should be confirmed by demonstrating inhibition in the presence of neutralizing anti-IFNγ antibodies.

3. Induction of Expression of MHC Class II Antigens

In this bioassay, IFNγ is quantitated based on its ability to induce expression of MHC class II antigens on appropriate target cells (the myelomonocytic cell line WEHI-3 for murine IFNγ assays, and the colon adenocarcinoma cell line, COLO-205 for human IFNγ assays).[127] Cells are cultured in the presence of IFNγ-containing samples or standards, and the expression of MHC class II molecules is then evaluated by flow cytometry.

Materials and Reagents

COLO-205 (ATCC Cat. No. CCL 1222) or WEHI-3 cells (ATCC Cat. No. TIB 68)

Recombinant human or murine IFNγ standards

Culture medium: RPMI-1640 containing 10% heat-inactivated FCS and supplemented with 2 mM L-glutamine, 10 mM HEPES, 100 U/ml penicillin, 100 µg/ml streptomycin, 50 µM 2-mercaptoethanol and 1 mM sodium pyruvate

Washing buffer: PBS containing 1% FCS (filter-sterilized)

Biotinylated antibodies specific for human HLA-DR (nonpolymorphic determinant) or murine I-Aᵈ (e.g., Pharmingen)

Avidin or streptavidin-fluorescein isothiocyanate (Av-FITC)

PBS-1% FCS containing 10 mM NaN₃

PBS-1% heat inactivated normal mouse serum containing 10 mM NaN₃

24-well tissue culture plates

12 x 75-mm centrifuge tubes

15- and 50-ml centrifuge tubes

Pasteur pipettes

Sterile rubber policeman

Refrigerated centrifuge (set at 4°C)
CO_2 incubator (humidified, set at 37°C and 5% CO_2)
Tabletop centrifuge
Reagents and equipment for flow cytometry (see Chapter 2)

Protocol:

(Note: All materials and reagents must be sterile and proper aseptic technique must be used when handling the cells. The protocol for cell staining does not require sterility.)

1. Harvest indicator cells and resuspend them in culture medium to a density of 5 x 10^5 cells/ml.
2. Dilute samples and standards (1 to 100 U/ml) in culture medium. Test four to five different dilutions. Add 0.5 ml of the diluted samples/standards to each well of a 24-well tissue culture plate. Add 0.5 ml of medium alone to two control wells (one for background staining and one for basal MHC class II expression).
3. Add 0.5 ml of medium and 1.0 ml of the indicator cell suspension per well. Incubate for 48 hours at 37°C and 5% CO_2.
4. Harvest the cells in each well by gently scraping with a rubber policeman and pipetteting the cell suspension up and down several times with a Pasteur pipette. Transfer the cells to a 12 x 75-mm centrifuge tubes and pellet the cells by centrifugation at 200 x g for 10 minutes at 4°C.
5. Aspirate the supernatants by suction and wash the cells twice with ice-cold PBS-1% FCS containing 10 mM NaN_3. After the last wash, resuspend the cells in 100 μl of PBS containing heat-inactivated normal mouse serum and 10 mM NaN_3. Incubate for 30 minutes on ice.
6. Add 20 μl of biotinylated anti-MHC class II antibody to each tube, except for the background staining control. Antibodies should be diluted in PBS-1% FCS containing 10 mM NaN_3 according to the manufacturer's recommendations or optimal staining concentrations should be determined beforehand. Normally, a final antibody concentration of 5 to 10 μg/ml gives adequate results.
7. Incubate the tubes for 60 minutes on ice. Wash twice with ice-cold PBS-1% FCS containing 10 mM NaN_3 and resuspend in 100 μl of the same buffer. Add avidin-FITC conjugate diluted according to manufacturer's recommendations.
8. Incubate for 30 minutes on ice. Wash twice with cold PBS-1% FCS containing 10 mM NaN_3 and resuspend in 0.5 ml of the same buffer. Maintain samples on ice until analysis.

9. Quantitate cellular fluorescence by flow cytometry as described in Chapter 2.
10. Construct a standard curve by plotting mean fluorescence intensity (MFI) as a function of the IFNγ concentration. Calculate IFNγ concentration in samples by reference to standard curve.

Comments:

1. Other indicator cell lines may also be used. IFNγ is species specific so the samples and the standards need to be assayed on indicator cells of the same species.
2. Antibodies directed against MHC class II (mouse or human) can be obtained from a variety of commercial sources. Alternatively, several hybridomas secreting anti-MHC class II antibodies are available from the ATCC.
3. If samples can not be analyzed on the same day, they can be fixed with 1% paraformaldehyde and kept at 4°C.

Table 1 summarizes the different indicator cell lines used for the cytokine bioassays, the species specificity of the assay and potentially interfering cytokines.

III. IMMUNOASSAYS

Under this description are assays that are based on the detection and quantitation of cytokines based on their reactivity with specific antibodies (monoclonal and/or polyclonal) or binding to specific cytokine receptors. This type of assays are fast becoming the preferred method for cytokine quantitation, mainly because of their speed and specificity. Unlike the bioassays, which depend on the measurement of the biological response of indicator cells, immunoassays are based on immunogenicity, and thus are inherently very specific, especially when monoclonal antibodies are used. The main advantages and disadvantages of immunoassays are listed below.

TABLE 1

Indicator Cell Lines Used in Cytokine Bioassays: Species Specificity and Interfering Cytokines

Assay	Indicator Cell	Species specificity Human	Mouse	Interfering cytokines
IL-1α/β	thymocytes (C3H/HeJ)	yes	yes	IL-2, IL-6, TNFα
	D10.G4.1	yes	yes	IL-2, IL-4
IL-2	HT-2, CTLL-2	yes	yes	IL-4, IL-15
IL-3	MC/9	no	yes	IL-4, SCF
	TF1	yes	no	IL-4, IL-5, IL-6, IL-13, GM-CSF, EPO
IL-4	HT-2, CTLL-2	no	yes	IL-2, IL-15
	CT.4S	no	yes	IL-2 (high conc.)
	CTLL-hu IL-4R	yes	yes	IL-2
IL-5	splenocytes + dextran-SO$_4$	?	yes	?
	BCL$_1$	yes	yes	IL-2
	TF1	yes	no	IL-3, IL-4, IL-6, IL-13 GM-CSF, EPO
IL-6	B9, 7TD1	yes	yes	IL-4 (high conc.)
IL-7	2E8	yes	yes	?
IL-8	human neutrophils	yes	yes	other chemokines
IL-9	M-07e	yes	yes	IL-3, GM-CSF
IL-10	Th1 clones (murine)	yes	yes	IL-4, TGFβ, IFNγ
	MC/9	yes	yes	IL-3, IL-4, SCF
IL-11	T1165, T10	yes	yes	IL-6
IL-12	PBMC (human)	yes	yes	IL-2, IL-4, IL-7, IFNγ

(TABLE 1 continued)

Assay	Indicator Cell	Species specificity Human Mouse		Interfering cytokines
IL-13	B9	yes	yes	IL-6, IL-4 (high conc.)
IL-15	CTLL-2	yes	yes	IL-2, IL-4
TNFα/ TNFβ	L929	yes	yes	?
IFNs (α/β/γ)	1929 + VSV WISH + VSV	no yes	yes no	TNFα/β ?
IFNγ	WEHI-279 WEHI-3 COLO-205	no no yes	yes yes no	? ? ?

This table indicates the cytokine bioassay, the indicator cell lines, whether the indicator cell line can be used to assay the human and/or murine cytokine, and cytokines that may potentially interfere with the bioassay, either through stimulation or inhibition. Note that species specificity needs to be taken into consideration with interfering cytokines (i.e., murine IL-4 interferes with the IL-2 bioassay using murine CTLL-2 cells; however, human IL-4 does not interfere, since human IL-4 is not active on murine cells). Refer to text for details.

Advantages:

1. Very specific (differentiate between cytokine subtypes) and reproducible;
2. Fast (results are normally obtained in 3 to 4 hours);
3. Not susceptible to potentiation or inhibition by other cytokines;
4. Generally less susceptible than bioassays to interference by cytokine antagonists or soluble cytokine receptors (depends on the antibodies used for capture and detection).

Disadvantages:

1. May not differentiate between biologically active and inactive cytokines;
2. Sometimes less sensitive than bioassays (~50 pg/ml);

3. Relatively expensive (especially if relying on commercially available kits).

TABLE 2

Monoclonal Anti-Mouse Cytokine Antibodies

Assay	Capture	Detection	Blocking
IL-2	JES6-1A12	JES6-5H4	S4B6
IL-3	MP2-8F8	MP2-43D11	MP2-8F8
IL-4	11B11, BVD4-1D11	BVD6-24G2 BVD4-1D11	11B11
IL-5	TRFK-5	TRFK-4	TRFK-5
IL-6	MP5-20F3	MP5-32C11	MP5-20F3
IL-10	JES5-2A5	SXC-1	JES5-2A5, SCX-1
GM-CSF	MP1-22E9	MP1-31G6	MP1-22E9
TNFα	MP6-XT22	MP6-XT3	MP6-XT3
IFNγ	R4-6A2	XMG1.2	R4-6A2, XMG1.2

These antibodies can be obtained from Pharmingen (San Diego, CA). Some hybridomas are also available through the ATCC. Additional monoclonal and polyclonal anti-cytokine antibodies are also available from other vendors. This table is intended to serve for reference purposes only and does not imply endorsement of any specific company by the authors.

Although several formats have been employed in the immunoassay of cytokines, perhaps the most common is the "sandwich type" ELISA, in which two different anti-cytokine antibodies (monoclonal and/or polyclonal) are used for capture and detection, respectively. Quantitation is then based on the spectrophotometric determination of enzymatic activity linked to the detecting antibody. Because the technique is basically the same for any cytokine, with exception of the antibodies, only a "generic" method will be described here. However, as properties (e.g., affinity, avidity) of different antibodies may

128

vary, the conditions for a particular assay may have to be optimized for best results.

TABLE 3

Monoclonal Anti-Human Cytokine Antibodies

Assay	Capture	Detection	Blocking
IL-1β	ILB1-H6	ILB1-H21, ILB1-H67	
IL-4	8D4-8	MP4-25D2	MP4-25D2
IL-5	TRFK-5	JES1-5A10	JES1-5A10
IL-7	BVD10-40F6	BVD10-11C10	BVD10-40F6
IL-10	JES3-9D7	JES3-12G8	JES3-9D7
IL-13	JES10-35G12	JES10-2E10	
GM-CSF	BVD2-23B6	BVD2-21C11	BVD2-23B6
TNFα	MAb1	MAb11	MAb1

Most of these monoclonal antibodies can be obtained from Pharmingen, with exception of the anti-IL-1β and anti-IL-13, which can be obtained from the ATCC. Additional monoclonal and polyclonal anti-cytokine antibodies are also available from other vendors. This table is intended to serve for reference purposes only and does not imply endorsement of any specific company by the authors.

Monoclonal and polyclonal anti-cytokine antibodies can be purchased in purified form from a variety of companies (see Table 2 and Table 3), or the hybridomas can be obtained from the ATCC, grown in the laboratory and the antibodies purified (see Chapter 8). For cytokine immunoassays, two different monoclonal antibodies or a monoclonal and a polyclonal antibody can be combined for use in an ELISA. Alternatively, appropriate monoclonal or polyclonal antibodies can also be used for neutralization purposes in bioassays. Companies such as Pharmingen (San Diego, CA) offer blocking antibodies suitable for cytokine-neutralization studies in tissue culture, and pairs of monoclonal antibodies suitable for use in ELISAs, with the advantage that the detection antibody is sold already biotinylated. Other companies, such as Genzyme (Cambridge, MA), sell combinations of monoclonal and polyclonal antibodies that can be used for neutralization and/or for ELISAs.

A. GENERAL PROTOCOL FOR CYTOKINE ELISA

Materials and Reagents

Enhanced protein-binding plates (e.g., Immulon-2, Dynatech, Chantilly, VA)
Multichannel pipettete (50 to 200 µl)
Parafilm^R
Refrigerator
Incubator (37°C)
Multiwell spectrophotometer (ELISA Reader)
Capture and detection anti-cytokine antibodies (detection antibody is biotinylated)
Avidin (or Streptavidin)-alkaline phosphatase (or peroxidase) conjugate (e.g., Jackson Immunoresearch, Sigma, etc.)
Substrate:
    a) p-nitrophenylphosphate tablets (Sigma Cat. No. N-2640) for alkaline phosphatase;
    b) 2,2'-Azino-bis(3-ethylbenzthiazoline-6-sulfonic acid) (ATBS, Sigma Cat. No. A-1888) for peroxidase
Coating buffer: 0.1 M $NaHCO_3$, pH 8.2
Blocking buffer: Phosphate-buffered saline containing 10% FCS (PBS-10% FCS) or PBS containing 1% bovine serum albumin and 2% normal rat serum
Dilution buffer: PBS-10% FCS or PBS containing 1% bovine serum albumin and 5% normal rat serum
Washing buffer: PBS containing 0.05% Tween 20 (PBS-Tween)
Substrate solution:
    a) for alkaline phosphatase, dissolve p-nitrophenyl phosphate tablets to a final concentration of 1 mg/ml in 20 mM diethanolamine-HCl buffer, pH 9.5 containing 1 mM $MgCl_2$;
    b) for peroxidase, dissolve ATBS to a final concentration of 0.3 mg/ml in 0.1 M citric acid, pH 4.35; then add 10 µl of 30% H_2O_2 per 10 ml of solution.
Stopping solution:
    a) 10 N NaOH for alkaline phosphatase;
    b) 16% SDS solution in 40% N,N-dimethyl formamide (SDS/DMF) for peroxidase.

1. Coating Plates with Capture Antibody

1. Dilute purified anti-cytokine antibody to 1 to 2 μg/ml in coating buffer.
2. Using enhanced protein-binding plates (e.g., Immulon-2, Dynatech), add 100 μl of the diluted antibody per well.
3. Seal the plate with Parafilm™ and incubate for six hours at room temperature or overnight at 4°C.

2. Blocking the Plates

1. Discard the antibody solution and wash the wells twice with PBS.
2. Add 200 μl/well of blocking buffer.
3. Seal the plate with Parafilm and incubate for a minimum of two hours at 37°C or overnight at 4°C.
4. Plates can be used immediately after completion of the blocking step, or stored at 4°C without removing the blocking solution. If stored for later use, it is recommended to place the sealed plates inside plastic bags (e.g., Ziplock™ bags) to minimize evaporation.

3. Samples/Standards

1. Determine the number of wells needed. It is advisable to assay each sample at least in duplicate or triplicate. Include extra wells for blanks and recombinant cytokine standards (6 to 7 different concentrations in the 10 to 2,000 pg/ml range). (See comment # 3.)
2. Add samples and standards (50 μl/well). Cover, and incubate for at least 2 hours at 37°C or overnight at 4°C. (See comment # 4.)
3. Wash four times with PBS-0.05% Tween 20 (Polyoxyethylene-sorbitan monolaureate).

4. Secondary (Detection) Antibody

1. Dilute the biotinylated secondary (detection) antibody to 1 to 2 μg/ml in dilution buffer. (See comment # 5.)
2. Add 60 μl of diluted antibody per well, cover, and incubate for at least one hour at 37°C.
3. Wash six times with PBS-0.05% Tween 20.

5. Avidin-Enzyme Conjugate

1. Dilute the avidin-alkaline phosphatase (or alternatively avidin-peroxidase) conjugate to the manufacturer's recommended dilution

131

(usually 1 to 2 µg/ml) using dilution buffer. (See comment # 6.)
2. Add 100 µl per well, cover, and incubate for at least 30 minutes at 37°C.
3. Wash six times with PBS-Tween 20.

6. Substrate

1. Prepare the appropriate substrate solution (p-nitrophenyl phosphate for alkaline phosphatase, ATBS-H_2O_2 for peroxidase) at least 5 minutes before use. Substrate solution must be at room temperature.
2. Add 100 µl per well. Allow to develop at room temperature for 10 to 60 minutes.
3. Stop the reaction by addition of stop reagent. (25 µl/well of 10 N NaOH for alkaline-phosphatase reactions, or 100 µl/well of SDS/DMF solution for peroxidase reactions).
4. Read at OD 405 nm.

7. Calculation

1. For manual calculation, construct a standard curve by plotting standard concentration (log scale) in pg/ml against OD_{405}. Calculate sample concentration by comparing sample OD_{405} with the standard curve.
2. A number of computer programs are available for computer-assisted calculation.

Comments:

1. In our laboratory, PBS has given satisfactory results as a coating buffer when using enhanced protein-binding plates.
2. In our experience, the use of 2 to 5% normal rat serum in both the blocking and diluting buffers has provided good results in minimizing background on cytokine ELISAs based on rat monoclonal antibodies.
3. Standards should be diluted in a buffer that closely matches the composition of the actual samples (e.g., culture media, serum, etc.).
4. We have determined that incubation of samples at 37°C rather than 4°C, minimizes interference by soluble cytokine receptors, probably due to faster dissociation rates at higher temperatures.
5. The detection antibody may either be purchased already biotinylated or biotinylated in the laboratory (see Chapter 2). Alternatively, the secondary antibody may be used without biotinylation, in which case a biotinylated or enzyme-conjugated third antibody directed against the second antibody (usually anti-immunoglobulin antibody) will be used.

6. Azide is a potent inhibitor of peroxidase, consequently, azide-free buffers should be used throughout if peroxidase conjugates are used for detection.
7. Monoclonal antibody pairs suitable for use in human and murine cytokine ELISAs are commercially available from PharMingen (San Diego, CA).
8. "Kits" for human and murine cytokine measurement are available from several vendors. Some of these are

Amersham	Intergen
Bio-Source	Oncoscience
Du Pont	Perseptive Diagnostics
Endogen	R & D Systems
Genzyme Diagnostics	T Cell Diagnostics
Immunotech	

(This list does not pretend to be all-inclusive.)

B. PROTOCOL FOR ENHANCED ELISA METHOD

Although the protocol described above for cytokine ELISAs should prove adequate for normal measurements, increased sensitivity may be preferred under some conditions. Increasing the sensitivity of an ELISA may involve addition of a third step or use of an amplification reaction. We next describe a protocol for such an amplification reaction originally described by Stanley et al.[128] This protocol is based on the amplification of an alkaline-phosphatase enzymatic reaction and results in approximately a 10 to 20-fold increase in sensitivity.

Substrate Reaction:

$$NADPH \xrightarrow{\text{Alkaline Phosphatase}} NADH + Pi$$

Amplification Reaction:

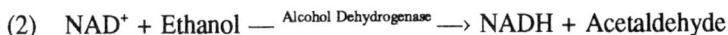

(1) $NADH + INT\text{-violet} \xrightarrow{\text{Diaphorase}} Formazan (red) + NAD^+$

(2) $NAD^+ + Ethanol \xrightarrow{\text{Alcohol Dehydrogenase}} NADH + Acetaldehyde$

Materials and Reagents

TBS: 50 mM Tris-HCl, 150 mM NaCl, pH 7.5
β-NADPH (reduced form, Sigma Cat. No. N-1630)
Alcohol Dehydrogenase (Sigma Cat. No. A-3263)
Diaphorase suspension (Sigma Cat. No. D-3752)
p-Iodonitrotetrazolium (INT-violet, Sigma Cat. No. I-8377)
Substrate solution: dissolve 2 mg of β-NADPH in 12 ml of 50 mM
diethanolamine-HCl buffer, pH 9.5; containing 1 mM MgCl$_2$
Amplification solution: first dissolve 2 mg of lyophilized alcohol
dehydrogenase in 11 ml of 50 mM phosphate buffer, pH 7.0, then add
60 µl of diaphorase suspension and 1 ml of a 6 mg/ml solution of p-
iodonitrotetrazolium violet in 36% ethanol. Filter.
Stopping solution: 0.3 M H$_2$SO$_4$

1. Follow sections 1 through 5 as described above for the general protocol for cytokine ELISA, using an avidin-alkaline phosphatase conjugate.
2. Wash plates four times with PBS-Tween 20 and six times with TBS in order to remove all phosphate.
3. Prepare the substrate solution at least 10 minutes before use. Substrate solution has to be at room temperature.
4. Add 50 µl of substrate solution per well. Incubate for 5 to 10 minutes at room temperature.
5. Prepare the amplifier solution at least 10 minutes before use and make sure solution is at room temperature.
6. Without removing the substrate solution from the wells, add 50 µl of the amplifier solution per well. Allow to develop for 5 to 10 minutes at room temperature.
7. Stop the reaction by addition of 50 µl per well of a 0.3 M H$_2$SO$_4$ solution.
8. Measure OD at 490 nm.

Comments:

1. Amplifier solution should be colorless or slightly yellow. Discard and prepare again if it develops a pink to red color. Unlike the substrate solution, the amplifier solution does not store well at -20°C and it's best prepared fresh.
2. An "ELISA Amplification System" kit based in the same reaction is available from GIBCO-BRL (Cat. No. 9589SA).

REFERENCES

1. Arai, K., Lee, F., Miyajima, A., Miyatake, S., Arai, N., and Yokota, T., Cytokines: coordinators of immune and inflammatory responses, *Annu. Rev. Biochem.*, 59, 783, 1990.
2. Oppenheim, J. J. and Sklatvala, J., Cytokines and their receptors, in *Clinical Applications of Cytokines: Role in Pathogenesis, Diagnosis and Therapy*, Oppenheim, J. J., Rossio, J. and Gearing, A., Eds., Oxford University Press, New York, 1993, 3.
3. Abbas, A. K., Lichtman, A. H., and Pober, J. S., *Cellular and Molecular Immunology, 2nd. Ed.*, Saunders, Philadelphia, 1994, 240.
4. Mosmann, T. R., Cherwinski, H., Bond, M. W., Giedlin, M. A., and Coffman, R. L., Two types of murine helper T cell clone. I. Definition according to profiles of lymphokine activities and secreted proteins, *J. Immunol.*, 136, 2348, 1986.
5. Mosmann, T. R. and Coffman, R. L., Th1 and Th2 cells: different patterns of lymphokine secretion lead to different functional properties, *Annu. Rev. Immunol.*, 7, 145, 1989.
6. Rossio, J. L., and Gearing, A. J. H., Measurement of cytokines, in *Clinical Applications of Cytokines: Role in Pathogenesis, Diagnosis and Therapy*, Oppenheim, J. J., Rossio, J. and Gearing, A., Eds., Oxford University Press, New York, 1993, 16.
7. Watson, J., Continuous proliferation of murine antigen-specific helper T lymphocytes in culture, *J. Exp. Med.*, 150, 1510, 1979.
8. Gillis, S., Ferm, M. M., Ou, W. and Smith, K. A., T cell growth factor: parameters of production and a quantitative microassay for activity, *J. Immunol.*, 120, 2027, 1978.
9. Fernandez-Botran, R., Krammer, P. H., Diamantstein, T., Uhr, J. W., and Vitetta, E.S., B cell-stimulatory factor 1 (BSF-1) promotes growth of helper T cell lines, *J. Exp. Med.*, 164, 580, 1986.
10. Grabstein, K. H., Eisenman, J., Shanebeck, K., Rauch, C., Srinivasan, S., Fung, V., Beers, C., Richardson, J., Schoenborn, M. A., Ahdieh, M., Johnson, L., Alderson, M. R., Watson, J. D., Anderson, D. M., and Giri, J. G., Cloning of a T cell growth factor that interacts with the β chain of the interleukin-2 receptor, *Science*, 264, 965, 1994.
11. Dinarello, C. A., Biology of interleukin-1, *FASEB J.*, 2, 108, 1988.
12. Kurt-Jones, E. A., Beller, D. I., Mizel, S. B., and Unanue, E. R., Identification of a membrane-associated interleukin-1 in macrophages, *Proc. Natl. Acad. Sci. U.S.A.*, 82, 1204, 1985.

13. Sims, J. E., Acres, R. B., Grubin, C. E., McMahan, C. J., Wignall, J. M., March, C. J., and Dower, S. K., Cloning the interleukin-1 receptor from human T cells, *Proc. Natl. Acad. Sci. U.S.A.*, 86, 8946, 1989.

14. McMahan, C. J., Slack, J. L., Mosley, B., Cosman, D., Lupton, S. D., Brunton, L. L., Grubin, C. E., Wignall, J. M., Jenkins, N. A., Branan, C. I., Copeland, N. G., Huebner, K., Croce, C. M., Cannizarro, L. A., Benjamin, D, Dower, S. K., Spriggs, M. K., and Sims, J. E., A novel IL-1 receptor, cloned from B cells by mammalian expression, is expressed in many cell types, *EMBO J.*, 10, 2821, 1991.

15. Hannum, C. H., Wilcox, C. J., Arend, W. P., Joslin, F. G., Dripps, D. J., Heimdal, P. L., Armes, L. G., Sommer, A., Eisenberg, S. P., and Thompson, R. C., Interleukin-1 receptor antagonist activity of a human interleukin-1 inhibitor, *Nature (London)*, 343, 336, 1990.

16. Gery, I., Gershon, R. K., and Waksman, B. H., Potentiation of the T lymphocyte response to mitogens. I. The responding cell, *J. Exp. Med.*, 136, 128, 1972.

17. Kaye, J., Gillis, S., Mizel, S. B., Shevach, E. M., Malek, T. R., Dinarello, C. A., Lachman, L. B., and Janeway, C. A., Jr., Growth of a cloned helper T cell line induced by a monoclonal antibody specific for the antigen receptor: Interleukin-1 is required for the expression of receptors for interleukin-2, *J. Immunol.*, 133, 1339, 1984.

18. Gillis, S., and Mizel, S. B., T cell lymphoma model for the analysis of interleukin 1-mediated T-cell activation, *Proc. Natl. Acad. Sci. U.S.A.*, 78, 1133, 1981.

19. Gearing, A. J., Bird, C. R., Bristow, A., Poole, S., and Thorpe, R., A simple sensitive bioassay for interleukin-1 which is unresponsive to 10^3 units of interleukin-2, *J. Immunol. Methods*, 99, 7, 1987.

20. Schmidt, J. A., Mizel, S. B., Cohen, D., and Green, I., Interleukin 1, a potential regulator of fibroblast proliferation, *J. Immunol.*, 128, 2177, 1982.

21. Lachman, L. B., Brown, D. C., and Dinarello, C. A., Growth promoting effect of recombinant interleukin 1 and tumor necrosis factor for a human astrocytoma cell line, *J. Immunol.*, 138, 2913, 1987.

22. Kaye, J., Porcelli, S., Tite, J., Jones, B., and Janeway, C. A., Jr., Both a monoclonal antibody and antisera specific for determinants unique to cloned helper T cell lines can substitute for antigen and antigen-presenting cells in the activation of T cells, *J. Exp. Med.*, 158, 836, 1983.

23. Smith, K. A., Interleukin-2: inception, impact and implications, *Science*, 240, 1169, 1988.

24. **Minami, Y., Kono, T., Miyazaki, T., and Taniguchi, T.,** The IL-2 receptor complex: its structure, function and target genes, *Annu. Rev. Immunol.,* 11, 245, 1993.

25. **Russell, S. M., Keegan, A. D., Harada, N., Nakamura, Y., Noguchi, M., Leland, P., Friedmann, M. C., Miyajima, A., Puri, R. K., Paul, W. E., and Leonard, W. J.,** The interleukin-2 receptor γ chain is a functional component of the interleukin-4 receptor, *Science,* 262, 1880, 1993.

26. **Kondo, M., Takeshita, T., Ishii, N., Nakamura, M., Watanabe, S., Arai, K., and Sugamura, K.,** Sharing of the interleukin-2 (IL-2) receptor γ chain between receptors for IL-2 and IL-4, *Science,* 262, 1874, 1993.

27. **Noguchi, M., Nakamura, Y., Russell, S. M., Ziegler, S. F., Tsang, M., Cao, X., and Leonard, W. J.,** Interleukin-2 receptor γ chain: a functional component of the interleukin-7 receptor, *Science,* 262, 1877, 1993.

28. **Zurawski, S. M., Vega, F., Jr., Huyghe, B., and Zurawski, G.,** Receptors for interleukin-13 and interleukin-4 are complex and share a novel component that functions in signal transduction, *EMBO J.,* 12, 2663, 1993.

29. **Rubin, L.A. and Nelson, D.L.,** The soluble interleukin-2 receptor: biology, function and clinical application, *Ann. Intern. Med.,* 113, 619, 1990.

30. **Bazill, G. W., Haynes, M., Garland, J., and Dexter, D. M.,** Characterization and partial purification of a hemopoietic growth factor in WEHI-3 conditioned medium, *Biochem. J.,* 210, 747, 1983.

31. **Ben-Sasson, Z., Le Gros, G., Conrad, D. H., Finkelman, F., and Paul, W. E.,** Cross-linking Fc receptors stimulate splenic non-B non-T cells to secrete IL-4 and other lymphokines, *Proc. Natl. Acad., Sci., U.S.A.,* 87, 1421, 1990.

32. **Schrader, J. W.,** The panspecific hemopoietin of activated T lymphocytes (interleukin-3), *Annu. Rev. Immunol.,* 4, 205, 1986.

33. **Nicola, N. A.,** Hematopoietic cell growth factors and their receptors, *Annu. Rev. Biochem.,* 58, 45, 1990.

34. **Nicola, N. A., and Metcalf, D.,** Subunit promiscuity among hemopoietic growth factor receptors, *Cell,* 67, 1, 1991.

35. **Tavernier, J., Devos, R., Cornelis, S., Tuypens, T., Van der Heyden, J., Fiers, W. and Plaetnick, G.,** A human high affinity interleukin-5 receptor (IL5R) is composed of an IL-5-specific α chain and a β chain shared with the receptor for GM-CSF, *Cell,* 66, 1175, 1991.

36. **Dexter, T. M., Garland, J., Scott, P., Scolnick, E., and Metcalf, D.,** Growth of factor dependent hemopoietic precursor cell lines, *J. Exp. Med.,* 152, 1036, 1980.

37. **Le Gros, G. S., Gillis, S., and Watson, J. D.,** Induction of IL-2 responsiveness in a murine IL-3-dependent T cell line, *J. Immunol.,* 139, 4009, 1985.

38. **Greenberger, J. S., Sakakeeny, M. A., Humphries, R., Eaves, C., and Echner, R.,** Demonstration of a permanent factor-dependent multipotential hemopoietic progenitor cell lines, *Proc. Natl. Acad. Sci. U.S.A.,* 80, 2931, 1983.

39. **Smith, C. A., and Rennick, D. M.,** Characterization of a murine lymphokine distinct from IL-2 and IL-3, possessing T cell growth factor activity and mast cell growth factor activity, *Proc. Natl. Acad. Sci. U.S.A.,* 83, 1857, 1986.

40. **Kitamura, T., Tauge, T., Terasawa, T, Chiba, S., Kuwaki, T., Miyagawa, K., Piao, Y.-F., Miyazono, K., Urabe, A., and Takaku, F.,** Establishment and characterization of a unique human cell line that proliferates dependently on GM-CSF, IL-3 or erythropoietin, *J. Cell Physiol.,* 140, 323, 1989.

41. **Le Gros, G. S., Le Gros, J. E., and Watson, J. D.,** The induction of lymphokine synthesis and cell growth in IL-3-dependent cell lines using antigen-antibody complexes, *J. Immunol.,* 139, 422, 1987.

42. **Paul, W.E.,** Interleukin-4: A prototypic immunoregulatory lymphokine, *Blood,* 7, 1627, 1991.

43. **Plaut, M., Pierce, J. H., Watson, D. J., Hanley-Hyde, J., Nordan, R. P., and Paul, W. E.,** Mast cell lines produce lymphokine in response to cross-linkage of FCεRI or to calcium ionophores, *Nature,* 339, 64, 1991.

44. **Vitetta, E. S., Fernandez-Botran, R., Myers, C. D., and Sanders, V. M.,** Cellular interactions in the humoral immune response, *Adv. Immunol.,* 45, 1, 1989.

45. **Finkelmann, F. D., Katona, I. M., Urban, J. F., Jr., Beckmann, M. P., Park, L. S., Schooley, K. A., Coffman, R. L., Mosmann, T. R., and Paul, W. E.,** Lymphokine control of in vivo immunoglobulin isotype selection, *Annu. Rev. Immunol.,* 8, 303, 1990.

46. **Swain, S.L., Weinberg, A. D., English, M., and Huston, G.,** IL-4 directs the development of Th2-like helper effectors, *J. Immunol.,* 145, 3796, 1990.

47. **Abeshira-Amar, O., Gilbert, M., Joliy, M., Theze, J., and Jankovic, D. L.,** IL-4 plays a dominant role in the differential development of Th0 into Th1 and Th2 cells, *J. Immunol.,* 148, 3820, 1992.

48. Seder, R. A., Paul, W. E., Davis, M. M., and de St Groth, B. F., The presence of interleukin-4 during in-vitro priming determines the lymphokine producing potential of CD4⁺ T cells from T cell receptor transgenic mice, *J. Exp. Med.,* 176, 1091, 1992.

49. Maher, D.W., Davis, I., Boyd, A.W. and Morstyn, G., Human interleukin-4: an immunomodulator with potential therapeutic applications, *Prog. Growth Factor Res.,* 3, 43, 1991.

50. Ohara, J. and W. E. Paul., Receptors for B-cell stimulatory factor-1 expressed on cells of hematopoietic lineage, *Nature,* 325, 537, 1987.

51. Park, L. S., Friend, D., Sassenfeld, H., and Urdal, D, Characterization of the high-affinity cell-surface receptor for murine B cell stimulating factor, *Proc. Natl. Acad. Sci. U.S.A.,* 84, 1669, 1987.

52. Lowenthal, J. W., Castle, B. E., Christiansen, J., Schreurs, J., Rennick, D., Arai, N., Hoy, P., Takebe, Y., and Howard, M., Expression of high-affinity receptors for murine interleukin-4 (BSF-1) on hemopoietic and nonhemopoietic cells, *J. Immunol.,* 140, 456, 1988.

53. Mosley, B., Beckman, M. P., March, C. J., Idzerda, R. L., Gimpel, S. D., VandenBos, T., Friend, D., Alpert, A., Anderson, J., Jackson, J., Wignall, J. M., Smith, C., Gallis, B., Sims, J. E., Urdal, D., Widmer, M. B., Cosman, D., and Park, L. S., The murine interleukin-4 receptor: molecular cloning and characterization of secreted and membrane bound forms, *Cell,* 59, 335, 1989.

54. Harada, N., Castle, B. E., Gorman, D. M., Itoh, N., Schreurs, J., Barrett, R. L., Howard, M., and Miyajima, A., Expression cloning of a cDNA encoding the murine interleukin 4 receptor based on ligand binding, *Proc. Natl. Acad. Sci. USA,* 87, 857, 1990.

55. Idzerda, R. L., March, C. J., Mosley, B., Lyman, S. D., Vandenbos, T., Gimpel, S. D., Din, W. S., Grabstein, K. H., Widmer, M. B., Park, L. S., Cosman, D., and Beckmann, M. P., Human interleukin 4 receptor confers biological responsiveness and defines a novel receptor superfamily, *J. Exp. Med.,* 171, 861, 1990.

56. Fernandez-Botran, R., and Vitetta, E.S., A soluble, high-affinity interleukin-4-binding protein is present in the biological fluids of mice, *Proc. Natl. Acad. Sci. U.S.A.,* 87, 4202, 1990.

57. Noelle, R., Krammer, P. H., Ohara, J., Uhr, J. W., and Vitetta, E. S., Increased expression of Ia antigens on resting B cells: An additional role for B cell growth factor, *Proc. Natl. Acad. Sci. U.S.A.,* 81, 6149, 1984.

58. Howard, M., Farrar, J., Hilfiker, M., Johnson, B., Takatsu, K., Hamaoka, T., and Paul, W. E., Identification of a T cell-derived B cell growth factor distinct from interleukin-2, *J. Exp. Med.,* 155, 914, 1982.

59. Vitetta, E. S., Ohara, J., Myers, C., Layton, J., Krammer, P. H., and Paul, W. E., Serological, biochemical and functional identity of B cell-stimulatory factor01 and B cell differentiation factor for IgG$_1$, *J. Exp. Med.*, 162, 1726, 1985.

60. Coffman, R. L., and Carty, J., A T cell activity that enhances polyclonal IgE production and its inhibition by interferon-γ, *J. Immunol.*, 136, 949, 1986.

61. Hu-Li, J., Ohara, J., Watson, C., Tsang, W., and Paul, W. E., Derivation of a T cell line that is highly responsive to Il-4 and Il-2 (CT.4R) and of an IL-2-hyporesponsive mutant of that line (CT.4S), *J. Immunol.*, 142, 800, 1989.

62. Idzerda, R. L., March, C. J., Mosley, B., Lyman, S. D., VandenBos, T., Gimpel, S. D., Din, W. S., Grabstein, K. H., Widmer, M. B., Park, L. S., Cosman, D., and Beckmann, M. P., Human interleukin-4 receptor confers biological responsiveness and defines a novel receptor superfamily, *J. Exp. Med.*, 171, 861, 1990.

63. Defrance, T., Vanbervliet, B., Aubry, J. P., Takebe, Y., Arai, N., Miyajima, A., Yokota, T., Lee, F., Arai, K., DeVries, J., and Banchereau, J., B cell growth-promoting activity of recombinant human interleukin-4, *J. Immunol.*, 139, 1135, 1987.

64. Rousset, F., de Waal Malefijt, R., Slierendregt, B., Aubry, J. P., Bonnefoy, J. Y., Defrance, T., Banchereau, J., and DeVries, J. E., Regulation of Fc-receptor for IgE (CD23) and class II MHC antigen expression of Burkitt's lymphoma cell lines by human interleukin-4, *J. Immunol.*, 140, 2625, 1988.

65. Sanderson, C. J., Warren, D. J., and Strath, M., Identification of a lymphokine that stimulates eosinophil differentiation in vitro. Its relationship to interleukin-3, and functional properties of eosinophils produced in cultures, *J. Exp. Med.*, 162, 60, 1985.

66. Campbell, H. D., Tucker, W. Q. J., Hort, Y., Martinson, M. E., Mayo, G., Clutterbuck, E. J., Sanderson, E. J., and Young, I. G., Molecular cloning, nucleotide sequence, and expression of the gene encoding human eosinophil differentiation factor (interleukin-5), *Proc. Natl. Acad. Sci. U.S.A.*, 84, 6629, 1987.

67. Swain, S. L., Howard, M., Kappler, J., Marrack, P., Watson, J., Booth, R., Wetzel, M., and Dutton, R. W., Evidence for two distinct classes of murine B cell growth factors with activities in different functional assays, *J. Exp. Med.*, 158, 822, 1983.

68. Brooks, K., Yuan, D., Uhr, J. W., Krammer, P. H., and Vitetta, E. S., Lymphokine-induced IgM secretion by clones of neoplastic B cells, *Nature (London)*, 302, 825, 1983.

69. Takatsu, K., Kikuchi, Y., Takahashi, T., Honjo, T., Matsumoto, M., Harada, N., Yamaguchi, N., and Tominaga, T., Interleukin-5, a T-cell-derived B-cell differentiation factor also induces cytotoxic T lymphocytes, *Proc. Natl. Acad. Sci. U.S.A.*, 84, 4234, 1987.

70. Swain, S. L., and Dutton, R. W., Production of a B cell growth-promoting activity, (DL)BCGF, from a cloned T cell line and its assay on the BCL_1 B cell tumor, *J. Exp., Med.*, 156, 1821, 1982.

71. Van Snick, J., Interleukin-6: an overview, *Annu. Rev. Immunol.*, 8, 253, 1990.

72. Kishimoto, T., and Hirano, T., Molecular regulation of B lymphocyte response, *Annu. Rev. Immunol.*, 6, 485, 1988.

73. Gearing, D. P., Comeau, M. R., Friend, D. J., Gimpel, S. D., Thut, C. J., McGourty, J., Brasher, K. K., King, J. A., Gillis, S., Mosley, B., Ziegler, S. F., and Cosman, D., The IL-6 signal transducer, gp130: An oncostatin M receptor and affinity converter for the LIF receptor, *Science*, 255, 1434, 1992.

74. Honda, M., Yamamoto, S., Chang, M., Yasukawa, K., Suzuki, H., Saito, T., Osugi, Y., Tokunaga, T. and Kishimoto, T., Human soluble IL-6-receptor: Its detection and enhanced release by HIV infection, *J. Immunol.*, 148, 2175, 1992.

75. Aarden, L. A., De Groot, E. R., Schaap, O. L., and Lansdorp, P. M., Production of hybridoma growth factor by human monocytes, *Eur. J. Immunol.*, 17, 1411, 1987.

76. Van Snick, J., Cayphas, S., Vink, A., Uyttenhove, C., Coulie, P., and Simpson, R., Purification and NH_2-terminal amino acid sequence of a new T cell-derived lymphokine with growth factor activity for B cell hybridomas, *Proc. Natl. Acad. Sci. U.S.A.*, 83, 9679, 1986.

77. Namen, A. E., Schmierer, A. S., March, C. J., Overell, R. W., Park, L. S., Urdal, D. L., and Mochizuki, D. Y., B cell precursor growth promoting activity. Purification and characterization of a growth factor active on lymphocyte precursors, *J. Exp. Med.*, 167, 988,1988.

78. Goodwin, R. G., Lupton, S., Schmierer, A., Hjerrild, K. J., Jerzy, R., Clevenger, W., Gillis, S., Cosman, D., and Namen, A. E., Human interleukin-7: Molecular cloning and growth factor activity on human and murine B-lineage cells, *Proc. Natl. Acad. Sci. U.S.A.*, 86, 302, 1989.

79. Chazen, G. D., Pereira, G. M. B., Le Gros, G., Gillis, S., and Shevach, E. M., Interleukin-7 is a T cell growth factor, *Proc. Natl. Acad. Sci. U.S.A.*, 86, 5923, 1989.

80. Okazaki, H., Ito, M., Sudo, T., Hattori, M., Kano, S., Katsura, Y., and Minato, N., IL-7 promotes thymocytes bearing α/β or γ/δ T cell receptors in vitro: Synergism with IL-2, *J. Immunol.*, 143, 2917, 1989.

81. Goodwin, R.G., Friend, D., Ziegler, S.F., Jerzy, R., Falk, B.A., Gimpel, S., Cosman, D., Dower, S., March, C.J., Namen, A.E. and Park, L.S., Cloning of the human and murine interleukin-7 receptors: Demonstration of a soluble form and homology to a new receptor superfamily, *Cell*, 60, 941, 1990.

82. Park, L. S., Friend, D. J., Schmierer, A. S., Dower, S. K., and Namen, A. E., Murine interleukin 7 receptor. Characterization of an IL-7 dependent cell line, *J. Immunol.*, 171, 1073, 1990.

83. Whitlock, C. A., and Witte, O. N., Long-term culture of B lymphocytes and their precursors from murine bone marrow, *Proc. Natl. Acad. Sci. U.S.A.*, 79, 3608, 1982.

84. Ishihara, K., Median, K., Hayashi, S., Pietrangeli, C., Namen, A. E., Miyazaki, K., and Kincade, P. W., Stromal-cell and cytokine-dependent lymphocyte clones which span the pre-B to B-cell transition, *Devel. Immunol.*, 2, 149, 1991.

85. Moore, K. W., O'Garra, A., de Waal Maleyft, R., Vieira, P., and Mosmann, T. R., Interleukin-10, *Annu. Rev. Immunol.*, 11, 165, 1993.

86. Fiorentino, D. F., Bond, M. W., and Mosmann, T. R., Two types of mouse T helper cell. IV. TH2 clones secrete a factor that inhibits cytokine synthesis by TH1 clones, *J. Exp. Med.*, 170, 2081, 1989.

87. Moore, K. W., Vieira, P., Fiorentino, D. F., Trounstine, M. L., Khan, T. a., and Mosmann, T. R., Homology of cytokine synthesis inhibitory factor (IL-10) to the Epstein-Barr virus gene BCRFI, *Science*, 248, 1230, 1990.

88. Thompson-Snipes, L., Dhar, V., Bond, M. W., Mosmann, T. R., Moore, K. W., and Rennick, D., Interleukin 10: A novel stimulatory factor for mast cells and their progenitors, *J. Exp. Med.*, 173, 507, 1991.

89. Kobayashi, M., Fitz, L., Ryan, M., Hewick, R. m., Clark, S. C., Chan, S., Loudon, R., Sherman, F., Perussia, B., and Trinchieri, G., Identification and purification of a natural killer cell stimulatory factor (NKSF), a cytokine with multiple biological effects on human lymphocytes, *J. Exp. Med.*, 170, 827, 1989.

90. Stern, A. S., Podlaski, F. J., Hulmes, J. D., Pan, Y-C. E., Quinn, P. M., Wolitzky, A. G., Familletti, P. c., Stremlo, D. l., Truitt, T., Chizzonite, R., and Gately, M. K., Purification to homogeneity and partial characterization of cytotoxic lymphocyte maturation factor from human B-lymphoblastoid cells, *Proc. Natl. Acad. Sci. U.S.A.*, 87, 6808, 1990.

91. Gately, M. K., Desai, B. B., Wolitzky, A. G., Quinn, P. M., Dwyer, C. m., Podlaski, F. J., Familletti, P. C., Sinigaglia, F., Chizzonite, R., Gubler, U., and Stern, A. S., Regulation of human lymphocyte proliferation by a heterodimeric cytokine, IL-12 (cytotoxic lymphocyte maturation factor), *J. Immunol.*, 147, 874, 1991.

92. Scott, P., IL-12: Initiation cytokine for cell-mediated immunity, *Science*, 260, 496, 1993.

93. Gately, M. K., and Chizzonite, R., Measurement of human and mouse interleukin-12, in *Current Protocols in Immunology*, Coligan, J. E., Kruisbeek, A. M., Margulies, D. H., Shevach, E. M., and Strober, W., Eds, Greene Publishing and Wiley Interscience, New York, 1991, 6.16.1.

94. Zurawski, G., and de Vries, J. E., Interleukin-13, an interleukin 4-like cytokine that acts on monocytes and B cells, but not T cells, *Immunol. Today*, 15, 19, 1994.

95. McKenzie, A, N. J., Culpepper, J. A., de Waal Maleyft, R., Briere, F., Punnonen, J., Aversa, G., Sato, A., Dang, W., Cocks, B. G., Menon, S., de Vries, J. E., Banchereau, J., and Zurawski, G., Interleukin-13, a novel T cell-derived cytokine that regulates human monocyte and B cell function, *Proc. Natl. Acad. Sci. U.S.A.*, 90, 3735, 1993.

96. de Waal Maleyft, R., Figdor, C., Huijbens, R., Mohan-Peterson, S., Bennet, B., Culpepper, J., Dang, W., Zurawski, G., and de Vries, J. E., Effects of IL-13 on phenotype, cytokine production, and cytotoxic function of human monocytes, *J. Immunol.*, 151, 6370, 1993.

97. Doherty, T. M., Kastelein, R., Menon, S., Andrade, S., and Coffman, R. L., Modulation of murine macrophage function by interleukin-13, *J. Immunol.*, 151, 7151, 1993.

98. McKenzie, A. N. J., and Zurawski, G., Measurement of interleukin-13, in *Current Protocols in Immunology*, Coligan, J. E., Kruisbeek, A. M., Margulies, D. H., Shevach, E. M., and Strober, W., Eds, Greene Publishing and Wiley Interscience, New York, 1991, 6.18.1.

99. Kundig, T. M., Schorle, H., Bachmann, M. F., Hengartner, H., Zinkernagel, R. M., and Horak, I., Immune responses in interleukin-2 deficient mice, *Science*, 262, 1059, 1993.

100. Oppenheim, J.J., Zachariae, C. O. C., Mukaida, N., and Matsushima, K., Properties of the novel supergene "intercrine" cytokine family, *Annu. Rev. Immunol.*, 9, 817, 1991.

101. Larsen, C. G., Anderson, A. O., Oppenheim, J. J., and Matsushima, K., Production of inteleukin-8 by human dermal fibroblasts and keratinocytes in response to interleukin 1 or tumor necrosis factor, *Immunology*, 68, 31, 1989.

102. **Harvath, L., Falk, W., and Leonard, E. J.,** Rapid quantitation of neutrophil chemotaxis: Use of a polyvinylpyrrolidone-free polycarbonate membrane in a multiwell assembly, *J. Immunol. Methods,* 37, 39, 1980.
103. **Van Snick, J., Goethals, A., Renauld, J-C., Van Roost, E., Uyttenhove, C., Rubira, M. R., Moritz, R. L., and Simpson, R. J.,** Cloning and characterization of a cDNA for a new mouse T cell growth factor (P40), *J. Exp. Med.,* 169, 363, 1989.
104. **Yang, Y-C., Ricciardi, S., Ciarletta, A., Calvetti, J., Kelleher, K., and Clark, S. C.,** Expression cloning of a cDNA encoding a novel human hematopoietic growth factor: Human homologue of murine T cell growth factor P40, *Blood,* 74, 1880, 1989.
105. **Donahue, R. E., Yang, Y-C., and Clark, S. C.,** Human P40 T cell growth factor (interleukin-9) supports erythroid colony formation, *Blood,* 75, 2271, 1990.
106. **Paul, S. R., Bennett, F., Calvetti, J. A., Kelleher, K., Wood, C. R., O'Hara, R. M., Leary, A. C., Sibley, B., Clark, S. C., Williams, D. A., and Yang, Y-C.,** Molecular cloning of a cDNA encoding interleukin-11, a novel stromal cell-derived lymphopoietic and hematopoietic cytokine, *Proc. Natl. Acad. Sci. U.S.A.,* 87, 7512, 1990.
107. **Quesniaux, V., Mayer, P., Liehl, E., Turner, K., Goldman, S. and Fagg, B.,** Review of a novel hematopoietic cytokine, interleukin-11, *Intl. Rev. Exp. Pathol.,* 34A, 205, 1993.
108. **Yin, T., Miyazawa, K., and Yang, Y-C.,** Characterization of interleukin-11 receptor and protein tyrosine phosphorylation induced by interleukin-11 in mouse 373-L1 cells, *J. Biol. Chem.,* 267, 8347, 1992.
109. **Bennett, F., Giannotti, Clark, S. C., and Turner, K. J.,** Measurment of human interleukin-11, in *Current Protocols in Immunology,* Coligan, J. E., Kruisbeek, A. M., Margulies, D. H., Shevach, E. M., and Strober, W., Eds, Greene Publishing and Wiley Interscience, New York, 1991, 6.15.1.
110. **Vassalli, P.,** The pathophysiology of tumor necrosis factors, *Annu. Rev. Immunol.,* 10, 411, 1992.
111. **Carswell, E. A., Old, L. J., Kassel, R. L., Green, S., Fiore, N., and Williamson, B.,** An endotoxin-induced serum factor that causes necrosis of tumors, *Proc. Natl. Acad. Sci. U.S.A.,* 72, 3666, 1975.
112. **Le, J. and Vilcek, J.,** Tumor necrosis factor and interleukin 1: Cytokines with multiple overlapping biological activities, *Lab. Invest.,* 56, 234, 1987.
113. **Engelmann, H., Aderka, D., Rubinstein, M., Rotman, D. and Wallach, D.,** A tumor necrosis factor-binding protein purified to homogeneity from human urine protects cells from tumor necrosis factor toxicity, *J. Biol. Chem.,* 264, 11974, 1989.

114. **Seckinger, P., Isaaz, S. and Dayer, J-M.**, Purification and biologic characterization of a specific tumor necrosis factor alpha inhibitor, *J. Biol. Chem.*, 264, 11966, 1989.

115. **Gatanaga, T., Hwang, C.D., Kohr, W., Cappuccini, F., Lucci, J.A. III, Jeffes, E.W. Lentz, R., Tomich, J., Yamamoto, R.S. and Granger, G.A.**, Purification and characterization of an inhibitor (soluble tumor necrosis factor receptor) for tumor necrosis factor and lymphotoxin obtained from the serum ultrafiltrates of human cancer patients. *Proc. Natl. Acad. Sci. USA.*, 87, 8781, 1990.

116. **Hogan, M. M., and Vogel, S. N.**, Production of TNF by rIFN-γ-primed C3H/HeJ (Lps^d) macrophages requires the presence of lipid-A-associated proteins, *J. Immunol.*, 141, 4196, 1988.

117. **Chen, A. R., McKinnon, K. P., and Koren, H. S.**, LPS stimulates fresh human monocytes to lyse actinomycin D-treated WEHI 164 target cells via increased secretion of a monokine similar to tumor necrosis factor, *J. Immunol.*, 135, 3979, 1985.

118. **Pestka, S., and Baron, S.**, Definition and classification of the interferons, *Methods Enzymol.*, 78, 3, 1981.

119. **Uze, G., Lutfalla, G., and Gresser, I.**, Genetic transfer of a functional human interferon α receptor into mouse cells: cloning and its expression of its cDNA, *Cell*, 60, 225, 1990.

120. **Aguet, M., Dembic, Z., and Merlin, G.**, Molecular cloning and expression of the human interferon γ receptor, *Cell*, 55, 273, 1988.

121. **Farrar, M. A., and Schreiber, R. D.**, The molecular cell biology of interferon-gamma and its receptor, *Annu. Rev. Immunol.*, 11, 571, 1993.

122. **Rubinstein, S., Familletti, P. C., and Pestka, S.**, Convenient assay for interferons, *J. Virol.*, 37, 755, 1981.

123. **King, D. P., and Jones, P. P.**, Induction of Ia and H-2 antigens on a macrophage cell line by murine interferon, *J. Immunol.*, 131, 315, 1983.

124. **Schreiber, R. d., Hicks, L. J., Celada, A., Buchmeier, N. A., and Gray, P. W.**, Monoclonal antibodies to murine gamma interferon which differentially modulate macrophage activation and antiviral activity, *J. Immunol.*, 134, 1609, 1985.

125. **Reynolds, D. S., Boom, W. H., and Abbas, A. K.**, Inhibition of B lymphocyte activation by interferon-γ, *J. Immunol.*, 139, 767, 1987.

126. **Familletti, P. C., Rubinstein, and Pestka, S.**, A convenient and rapid cytopathic effect inhibition assay for interferon, *Methods Enzymol.*, 78, 387, 1981.

127. **Schreiber, R. D.**, Measurement of mouse and human IFNγ, in *Current Protocols in Immunology*, Coligan, J. E., Kruisbeek, A. M., Margulies, D. H., Shevach, E. M., and Strober, W., Eds, Greene Publishing and Wiley Interscience, New York, 1991, 6.8.1.

128. **Stanley, C. J., Johannsson, A. and Self, C. H.**, Enzyme amplification can enhance both the speed and the sensitivity of immunoassays, *J. Immunol. Methods*, 83, 89, 1985.

Chapter 7

CYTOKINE RECEPTORS

I. INTRODUCTION

Cytokines exert their effects through binding to cytokine receptors expressed on the membrane of their target cells. In fact, the expression of cytokine receptors is one major mechanism regulating cytokine activity *in vivo*.[1,2] Initial attempts to characterize cytokine receptors by radioreceptor assays and receptor cross-linking studies yielded important information regarding receptor affinity and numbers, and the physical characteristics of cytokine-binding receptor components. However, it was not until recent years, with the development of anti-cytokine receptor monoclonal antibodies and the use of molecular techniques for the cloning and sequencing of the genes encoding cytokine receptors, that a much clearer picture of cytokine receptors and their mechanisms of function has emerged.[2,3]

It is now known that, in general, functional cytokine receptors are composed of two or more different subunits, with one or two subunits contributing the ligand-binding function, and a different subunit functioning in signal-transduction.[3-8] It appears also that whereas the ligand-binding subunits alone have relatively low affinities for their ligands (low-affinity receptors), the interaction with the signal-transducing subunit stabilizes the binding, increasing the binding affinity several orders of magnitude (high-affinity receptors). Thus, these findings have provided an explanation for the previous observations of the heterogeneity of cytokine-binding sites (high- and low-affinity cytokine receptors) and that the biological activity was associated with the high-affinity receptors.[3-8]

Another observation that has found explanation by recent studies is that of the overlapping functions of obviously different cytokines. It is now evident that different cytokine receptors "share" common signal-transducing subunits, such as the case of the IL-2R γ-chain among IL-2R, IL-4R, IL-7R, IL-13R and IL-15R;[3-6,9] and the IL-3R β-chain shared among IL-3R, IL-5R and GM-CSF-R.[7] Moreover, sequence homology comparisons have indicated that many cytokine receptors share common structural characteristics.[3,10] Based on these properties, cytokine receptors belong to one of five different "cytokine receptor families":[11]

1) The hemopoietin receptor family (including IL-2Rβ and γ chains, IL-4R, IL-5R, IL-7R, and many more cytokine and hormone receptors);

2) The TNF receptor family (including TNF receptors type I and II, CD40, nerve growth factor receptor [NGF-R]);

3) The immunoglobulin superfamily (including IL-1 receptors type I, IL-6R α-chain);

4) The IFN receptor family (including IFNα/β-R and IFNγ-R);

5) The seven-transmembrane helix family (including receptors for IL-8 and other chemokines).

Although membrane receptors have the obvious function of ligand binding and signal transduction, mounting evidence has indicated that many cytokine receptors also exist in soluble form.[1] These are truncated counterparts of the membrane receptors, generated either by "receptor shedding" or encoded by separate mRNAs, that are released into biological fluids. The physiologic role of these soluble cytokine receptors (sCR) is not yet clear; however, since they retain their ligand-binding properties and inhibit binding of their respective ligands to membrane receptors, they are usually regarded as "cytokine antagonists".[1] Nonetheless, it has been demonstrated by several studies, that sCR can, under some conditions, potentiate the activity of their ligands *in vivo*, probably as a result of increased half-life and/or altered biodistribution.[1,12,13] The functional role of sCR in immune responses and their potential use as prognostic "markers" remains to be studied further.

There are several approaches to the study of cytokine receptors, depending on the information sought. For example:

1) Radioreceptor assays, which measure binding of labeled cytokines to their receptors on target cells, give important information of receptor number and affinity(ies). These assays are very sensitive, but usually require relatively large numbers of cells and adequate supply of cytokine (for labeling and competition studies).

2) Receptor cross-linking studies, which analyze the structure of cytokine-binding subunits by covalent cross-linking to the radiolabeled cytokine.

3) Flow cytometry, detecting cytokine receptor expression on cells based on reactivity with fluorochrome-labeled anti-receptor antibodies or cytokines. These assays require significantly lower number of cells that the radioreceptor assays, but their sensitivity is lower. Differences in the levels of receptor expression can be evaluated in a mixed cell population.

4) Enzyme-linked immunoassays are the method of choice for the detection and quantitation of soluble cytokine receptors. The basic technique is the same as the one described for the cytokine ELISAs, with the only difference of the specific anti-cytokine antibodies. In addition, cellular ELISAs could also be adapted for the quantitation of membrane receptors.

A. RADIORECEPTOR ASSAY

The protocol described here allows the determination of receptor number and dissociation rate(s) based on Scatchard plots of equilibrium binding data.[14] The target cells are incubated in the presence of several concentrations of radiolabeled cytokine for a length of time enough to allow binding equilibrium to take place (~30 to 60 minutes). The amount of cytokine bound at each concentration is determined by centrifuging the cells through an oil gradient, allowing rapid separation of cell-bound from free cytokine, and then counting the cell pellets. Based on a plot of the amount of cytokine specifically bound versus the ratio of bound/free cytokine at each concentration, it is possible to determine whether a homogeneous (straight line) or heterogenous class of receptors exist (curvilinear line). In addition, receptor number and dissociation constant (Kd) can be estimated for each class of receptors. Readers are encouraged to familiarize themselves with the theoretical bases for these assays by consulting appropriate sources,[15,16] as they will not be discussed in this chapter.

1. Binding Assay

Materials and Reagents

Target cells (cell population to be assayed)
Culture medium (RPMI-1640 containing 10% FCS)
PBS containing 1% FCS
Radiolabeled cytokine (either labeled in the laboratory or purchased commercially, e.g., Amersham, Du Pont NEN)
Recombinant cytokine
Dibutyl phthalate/dioctyl phthalate oil mixture (45:55) (Sigma)
Microcentrifuge tubes (0.5- or 1.5-ml Eppendorf tubes)
400-µl microcentrifuge tubes with tips for collection of cell pellets (Sarstedt Cat. No. 72.702)
12 x 75-mm plastic tubes
Ice bath
Tabletop centrifuge, 4°C

Microcentrifuge (Eppendorf)
Large scissors or nail clippers
γ-counter

Protocol:

1. First, determine the range and number of dilutions to be assayed. The range should be well below and above the dissociation rate (Kd). Since most cytokine receptors have Kds on the range of 10 to 100 pM, a range covering from 0.5 pM to 10 nM should be attempted. Allow four tubes for each dilution. Two will receive the labeled cytokine only (for the estimation of total binding), and the other two will receive same concentration of labeled cytokine plus a 200-fold molar excess of unlabeled cytokine (for the estimation of nonspecific binding). Specific binding will be then calculated by subtraction of nonspecific from total binding at each concentration.

2. Binding reactions are performed in a final volume of 200 μl. Dilute the radiolabeled cytokine using culture medium and add 50 μl per tube (Note that there will be a 4-fold dilution once the cells and other reagents are added, so make your dilutions accordingly.) Dilute the unlabeled cytokine stock and add 50 μl to each of the two tubes for the measurement of nonspecific binding, again taking into account the 1:4 dilution, or 50 μl of culture medium to the other two tubes measuring the total binding. Keep the tubes on an ice bath throughout the procedure.

3. Harvest the target cells and wash three times with PBS-1% FCS by centrifugation at 200 x g. Count and resuspend the cells in culture medium to a density of 1×10^7 cells/ml. Using 1.5-ml Eppendorf tubes, add 100 μl of the cell suspension per tube (this will give a final cell number of 1×10^6 cells per tube). Depending on the availability of cells and the relative number of receptors per cell, the number of cells per reaction can be varied from 0.5 to 5×10^6 cells per tube. Incubate on ice with gentle shaking until equilibrium is reached. The time to reach binding equilibrium needs to be determined in initial experiments, as it can vary from minutes to several hours.

4. While the Eppendorf tubes containing the cells and the labeled cytokine are being incubated, pipettete 180 μl of the phthalate oil mixture into the same number of 400-μl microcentrifuge tubes. Spin the tubes briefly (1 to 2 seconds) in an Eppendorf microcentrifuge (10,000 rpm) in order to release air bubbles trapped at the bottom of the tubes.

5. After equilibrium has been reached, load the contents of each Eppendorf tube on top of their respective oil-containing tubes.

Centrifuge the 400-µl tubes in an Eppendorf microcentrifuge (10,000 rpm) for 20 seconds. The cells should be pelleted at the bottom of the tips, whereas the medium should remain on top of the oil. With the aid of a large scissors or nail clippers, cut the tips of the tubes and place them in 12 x 75-mm plastic tubes. Care should be taken not to allow the oil to drip into these tubes.

6. Count the tips in a γ-counter. Also, determine the total cpm added per tube by counting 20 µl of each labeled cytokine dilution.

7. For calculations see section below.

Comments:

1. See below for protocol for iodination (labeling) of cytokines.

2. When studying binding of a cytokine to cells that secrete it or that have been previously exposed to it (for example, binding of IL-2 to HT-2 cells grown in IL-2-containing medium), cells must be first exposed to acid in order to remove any cytokine already bound to the membrane receptors. For this, first wash the cells in PBS-1% FCS to remove the free cytokine, then pellet the cells by centrifugation and resuspend the pellet for 30 seconds in 1 to 2 ml of culture medium adjusted to pH 3.0 with HCl. Quickly add the cells to 20 ml of cold, regular culture medium and centrifuge. Wash the cells once and resuspend in culture medium at 4°C.

2. Iodination Protocol

This protocol is based on the procedure described by Lowenthal et al.,[17] using Iodogen (Pierce)-coated tubes. We have used it with good results for the iodination of several cytokines, including rIL-4 and rIL-2. A number of alternative products for iodination are also available, such as Enzymobeads (Bio-Rad), Iodobeads (Pierce), Bolton-Hunter reagent (Amersham, Du Pont). In addition, [125I]-labeled human and mouse cytokines are also available from radioactive product vendors such as Amersham, Du Pont, ICN, etc.

Materials and Methods

Iodo-gen™ (Pierce, Cat. No. 28600)
Recombinant cytokine (carrier free)
1.5-ml Eppendorf tubes
Iodine-125 (NaI-carrier free) (Du Pont, Cat. No. NEZ-033L)
0.2 M borate buffer, pH 8.5
0.2 M HEPES with 0.3 M NaCl, pH 5.35

PBS-1% FCS
Affinity column with antibody specific for cytokine
Saline solution (0.15 M NaCl)
Elution buffer: 50 mM citric acid, 0.15 M NaCl, pH 2.0
1M Tris-HCl, pH 9.0
Carrier protein: 10 mg/ml BSA in PBS
Chloroform
γ-counter
Minicolumns (2 ml) (Bio-Rad)

Protocol:

1. Prepare a 1 mg/ml solution of Iodo-gen™ in chloroform. Further dilute to 40 µg/ml by adding 40 µl of this solution to 960 µl of chloroform. Place 25 µl/tube of this solution into 1.5-ml Eppendorf tubes (1 µg Iodo-gen™ per tube). Evaporate the solvent under a gentle stream of nitrogen. Cap the tubes and store in the dark at room temperature. The tubes can be stored for several months.

2. Add 1 to 5 µg of recombinant cytokine in a volume of 25 µl of buffer into the Iodogen-coated tube. Add approximately 1 mCi of ^{125}iodine diluted in 25 µl of the borate buffer, pH 8.5. Incubate on ice for 10 to 15 minutes.

3. Stop the reaction by transferring the contents to a clean Eppendorf tube. Add 50 µl of the HEPES-NaCl buffer and add to another Eppendorf tube containing approximately 1 ml of a 50% suspension of Sepharose beads coupled to the appropriate anti-cytokine monoclonal antibody in PBS-1% FCS.

4. Incubate for at least one hour at 4°C. Transfer the contents of the tube to a minicolumn (Bio-Rad) and wash beads with 15 x 1-ml aliquots of PBS-1% FCS, followed by 5 x 1-ml washes with saline. Monitor radioactivity.

5. Elute the bound material with 20 x 125-µl aliquots of elution buffer. Collect every two drops (~ 125 µl) into a separate tube containing 40 µl Tris-HCl, pH 9.0 and 10 µl BSA (10 mg/ml). Count radioactivity (it is best to take a 2-µl aliquot from each tube).

6. Plot cpm vs. fraction number. Pool fractions with peak of radioactivity. Aliquot and store at -70°C.

7. Determine the concentration of bioactive cytokine in the labeled preparation by an appropriate bioassay (see Chapter 6).

Comments:

1. The iodination protocol should be performed in a fume hood designated specifically for radioactive work. Follow appropriate precautions and safe handling of radioactive material.
2. Affinity chromatography is the most convenient way to remove free iodine. Alternatively, this separation could also be accomplished by gel filtration using a 1-ml column of Sephadex G-25. The affinity column medium can be prepared by coupling the appropriate anti-cytokine monoclonal antibody to Affi-gel 10 beads (Bio-Rad) following the manufacturer's instructions.
3. Labeled cytokines should be stored in the presence of a carrier protein.

3. Calculations

Scatchard plots of equilibrium binding data (concentration of bound ligand on the x axis vs. ratio of bound/free ligand at each concentration tested on the y axis) provide important information on several parameters, including:

a) Presence of a single or multiple classes of binding sites (e.g., high- and low-affinity receptors). Scatchard plots yielding a straight line are indicative of a single affinity, whereas a curved line may mean receptors of two or more different affinities and/or negative cooperativity.
b) Dissociation constant(s) (Kd), given by the reciprocal of the slope of the curve(s).
c) Number of binding sites (Bmax), given by the intercept of the line at the x axis.

These parameters can be calculated manually, or alternatively, with the aid of a number of computer programs for analysis of binding data.[18] These programs will calculate the best fit for the curves, and provide information about the classes and number of binding sites and their dissociation constants.

1. For calculation, create a table with the following columns:

a) The concentrations of [^{125}I]cytokine tested (in molar units).

b) Total cpm bound at each concentration.

c) Nonspecific binding at each concentration (cpm)

2. Calculate the specific binding at each concentration by subtracting nonspecific binding (column **c**) from total binding (column **b**). Nonspecific binding should be linear. Place results in column (**d**).

3. Write the total cpm added at each concentration in column (**e**). Now, calculate the concentration of bound ligand by dividing the values in column (**d**) [specific binding, in cpm] by the total cpm values (in column **e**) and multiplying the result by the ligand concentration (in column **a**). Write the result in column (**f**); this is the concentration of bound ligand in molar units.

$$[\text{Bound ligand}] \ (\mathbf{f}) \ = \ (\mathbf{d} \div \mathbf{e}) \ \text{x} \ \mathbf{a}$$

4. Calculate the free ligand concentration by subtracting the bound ligand concentration on column (**f**) from initial ligand concentration in column (**a**). Place the result in column (**g**) on the y axis

5. Calculate the ratio of bound/free ligand, by dividing the bound ligand concentration in column (**d**) by the free ligand concentration of column (**g**). Place the result in column (**h**). All the concentration values are in molar units.

6. Plot the bound ligand concentration, in column (**d**) on the x axis versus the bound/free ratio in column (**h**).

7. Calculate the best-fitting curve. If the resulting curve is a straight line (which can be fitted by simple linear regression analysis), this indicates a single affinity. The apparent dissociation constant is given by the reciprocal of the slope of the curve. Its values are in molar units. If the resulting curve is not straight, indicative of receptors of more than one affinity, the data can be resolved into two different affinity components (high- and low-affinity) which are plotted and analyzed separately, treating each component as a single class of receptors.

Dissociation constant (Kd) = 1/slope **(in molar units)**

8. From the linear regression analysis: $y = mx + b$, where "**m**" is the slope and "**b**" is the intercept at the y axis. In order to calculate the number of receptors per cell, calculate first the intercept at the x axis (maximal amount of ligand bound) by dividing the intercept at y (**b**) by the slope (**m**). The result (in molar units) is then multiplied by the assay volume in liters (2×10^{-4}), multiplied by 6.02×10^{23} molecules/mole (Avogadro's number) and divided by the number of cells used in each tube (e.g., 1×10^6 cells). Assuming that each receptor binds one single molecule of ligand, the resulting number should equal the number of receptors per cell.

154

$$\text{\# receptors/cell} = (\ b \div m)\ x\ \text{vol. in liters}\ x\ 6.02\ x\ 10^{23}\ x\ 1/\text{No. of cells}$$

B. CROSS-LINKING OF LABELED CYTOKINES TO THEIR RECEPTORS

Important information regarding the composition of high- and low-affinity receptors and the molecular weight of receptor subunits can be obtained by the covalent cross-linking of an iodinated cytokine to its membrane receptors followed by SDS-PAGE analysis of the cell lysate. Although a wide variety of homofunctional and heterofunctional cross-linking reagents are available (Pierce, Rockford, IL), two of the most commonly used in cytokine receptor analysis are disuccinimidyl suberate (DSS) and dithiobis(succinimidyl propionate) [DSP].[19-22] Cross-linking by the former reagent is irreversible, whereas cross-linking by DSP is reversible by sulfhydryl reagents. The basic protocol is similar.

Materials and Reagents

Target cells
Culture medium (RPMI 1640-10% FCS)
PBS-1% FCS
PBS
[^{125}I]labeled cytokine (see previous section)
Disuccinimidyl suberate (DSS) (Pierce Cat. No. 21555 G) or dithiobis(succinimidyl propionate) (DSP, Pierce Cat. No. 22585G)
Dimethylsulfoxide (DMSO)
PBS-containing 20 mM L-lysine, pH 7.25
Lysis buffer: PBS containing 1% NP 40, 1 mM phenylmethylsulfonyl fluoride and 1 mM EDTA, pH 7.25 (Sigma)
Reagents and equipment necessary for SDS-PAGE analysis and autoradiography
Centrifuge tubes (15 and 50 ml)
Eppendorf microcentrifuge tubes (1.5 ml)
Refrigerated centrifuge
Ice bath

Protocol:

1. Harvest the target cells. Wash the cells three times in cold PBS-1% FCS. Count and resuspend at $1\ x\ 10^7$ cells/ml in culture medium (15-ml centrifuge tube). Place the tube on ice.

2. Add the radiolabeled cytokine to the desired concentration. Most high-affinity cytokines receptors will be occupied if the ligand concentration is approximately 1 to 2 x 10⁻¹⁰ M. However, for binding to low-affinity receptors, higher concentrations of ligand (~ 1 to 2 x 10⁻⁹) are required. Incubate for 60 minutes on ice with occasional shaking.

3. Wash the cells once in 15 ml of ice-cold PBS and aspirate the supernatant. Resuspend cell pellet in 0.5 ml of cold PBS.

4. Dissolve DSP in DMSO to a final concentration of 100 mM. Make a 1:100 dilution by taking 10 μl of this solution and adding it to 990 μl cold PBS, while vortexing. Add 0.5 ml of the diluted DSP solution to the cell suspension. (Final DSP concentration is 0.5 mM, and final DMSO concentration is 0.5%.) For DSS, use a 1:40 dilution of a 100 mM DSS solution in DMSO (final concentration of DSS is 2 mM). Incubate for 30 minutes on ice with occasional shaking.

5. Inactivate and remove the cross-linker by washing once with 13 ml of cold PBS-containing L-lysine and twice with PBS-1% FCS.

6. Aspirate the last wash and resuspend the cell pellet in ~150 μl of lysis buffer. Incubate for 15 minutes on ice.

7. Transfer the lysate to a 1.5-ml microcentrifuge tube and centrifuge in an Eppendorf centrifuge (~10,000 rpm) for 5 minutes in order to separate nuclei and insoluble debris.

8. At this point, the supernatants can be mixed with an equal volume of 2X SDS-PAGE sample buffer and electrophoresed directly, or the lysates can be first immunoprecipitated with the appropriate anti-cytokine receptor antibodies and then analyzed by SDS-PAGE.

9. After electrophoresis, fix the gels with 40% methanol containing 10% trichloroacetic acid, dry the gel and visualize bands by autoradiography.

Comments:

1. Incubation of cells with ligand at 4°C prevents internalization of the receptor.

2. It is recommended to include specificity controls, in which the cells are incubated with the iodinated cytokine in the presence of a 100-fold molar excess of unlabeled cytokine. In addition, other controls should include:
   a) cells plus labeled cytokine without cross-linking reagent; and
   b) labeled cytokine alone plus cross-linking reagent.

3. The cross-linking reagent must be prepared fresh just prior to use.

4. Samples cross-linked with DSS can be analyzed by SDS-PAGE under both reducing and nonreducing conditions. Samples cross-linked with

DSP can be analyzed only under nonreducing conditions, as reducing agents cleave the cross-linker.

C.   ANALYSIS OF CYTOKINE RECEPTORS BY FLOW CYTOMETRY

The relative expression of cytokine receptors (or cytokine receptor subunits) can also be analyzed by flow cytometry after staining of target cells with appropriate fluorochrome-conjugated anti-cytokine receptor antibodies or cytokines. The advantages of this technique are that relatively small numbers of cells are needed and that no radioactive reagents are required. This method, however, is less sensitive and quantitative than the radioreceptor binding assay (cells expressing only a few hundred receptors per cell may not stain very brightly). Anti-cytokine receptor antibodies (both human and murine) are available from several sources and can be used directly labeled with FITC (or other fluorochrome) or coupled to a biotin-avidin-FITC detection system. (The protocols used for staining and analysis have been described in Chapter 2.)

In addition, both human and mouse cytokines coupled to FITC can be purchased from R & D International.

D.   ELISA FOR DETECTION OF SOLUBLE CYTOKINE RECEPTORS

Detection and quantitation of soluble cytokine receptors in tissue culture supernatants or biological fluids are usually performed by the same sandwich-ELISA method described in Chapters 6 and 9. The protocols for plate coating, capture and detection are identical, with exception of the appropriate anti-cytokine receptor antibodies. One important point to take into consideration when assaying soluble cytokine receptors is the fact that the binding of anti-receptor antibodies may be blocked by the presence of endogenously-secreted receptor ligand (cytokine). For example, we have determined that the reactivity of the murine soluble IL-4 receptor with M1, an anti-IL-4R monoclonal antibody is inhibited by the presence of IL-4. One approach to avoid interference by endogenously secreted IL-4 in this case is to pre-incubate the samples with an excess of a neutralizing anti-IL-4 monoclonal antibody (10 µg/ml) for 60 minutes at 37°C, and then performing the sIL-4R capture and detection.

Alternatively, detection of captured soluble cytokine receptors has been accomplished by binding of a radiolabeled cytokine preparation instead of a second anti-receptor monoclonal antibody. In this case, however, the capture antibody must not prevent ligand binding.

For soluble receptors with potential clinical diagnostic or prognostic applications, such as soluble IL-2R and soluble TNF-R, ELISA kits have been developed and can be purchased from different vendors, including AMAC, Immunotech, T Cell Diagnostics, and Biosource.

REFERENCES

1. **Fernandez-Botran, R.,** Soluble cytokine receptors: their role in immunoregulation, *FASEB J.,* 5, 2567, 1991.
2. **Taga, T., and Kishimoto, T.,** Cytokine receptors and signal transduction, *FASEB J.,* 7, 3387, 1993.
3. **Boulay, J. L., and Paul, W. E.,** The interleukin-4-related lymphokines and their binding to hemopoietin receptors, *J. Biol. Chem.,* 267, 20525, 1992.
4. **Kondo, M., Takeshita, T., Ishii, N., Nakamura, M., Watanabe, S., Arai, K., and Sugamura, K.,** Sharing of the interleukin-2 (IL-2) receptor γ chain between receptors for IL-2 and IL-4, *Science,* 262, 1874, 1993.
5. **Noguchi, M., Nakamura, Y., Russell, S. M., Ziegler, S. F., Tsang, M., Cao, X., and Leonard, W. J.,** Interleukin-2 receptor γ chain: a functional component of the interleukin-7 receptor, *Science,* 262, 1877, 1993.
6. **Zurawski, S. M., Vega, F., Jr., Huyghe, B., and Zurawski, G.,** Receptors for interleukin-13 and interleukin-4 are complex and share a novel component that functions in signal transduction, *EMBO J.,* 12, 2663, 1993.
7. **Nicola, N. A., and Metcalf, D.,** Subunit promiscuity among hemopoietic growth factor receptors, *Cell,* 67, 1, 1991.
8. **Gearing, D. P., Comeau, M. R., Friend, D. J., Gimpel, S. D., Thut, C. J., McGourty, J., Brasher, K. K., King, J. A., Gillis, S., Mosley, B., Ziegler, S. F., and Cosman, D.,** The IL-6 signal transducer, gp130: An oncostatin M receptor and affinity converter for the LIF receptor, *Science,* 255, 1434, 1992.
9. **Grabstein, K. H., Eisenman, J., Shanebeck, K., Rauch, C., Srinivasan, S., Fung, V., Beers, C., Richardson, J., Schoenborn, M. A., Ahdieh, M., Johnson, L., Alderson, M. R., Watson, J. D., Anderson, D. M., and Giri, J. G.,** Cloning of a T cell growth factor that interacts with the β chain of the interleukin-2 receptor, *Science,* 264, 965, 1994.

10. **Bazan, J. F.,** A novel family of growth factor receptors: A common binding domain in the growth hormone, prolactin, erythropoietin and Il-6 receptors, and the p75 IL-2 receptor β-chain, *Biochem. Biophys. Res. Commun.,* 164, 788, 1989.

11. **Abbas, A. K., Lichtman, A. H., and Pober, J. S.,** *Cellular and Molecular Immunology, 2nd. Ed.,* Saunders, Philadelphia, 1994, 240.

12. **Sato, T. A., Widmer, M. B., Finkelman, F. D., Madani, H., Jacobs, C. A., Grabstein, K. H., and Maliszewski, C. R.,** Recombinant soluble murine IL-4 receptor can inhibit or enhance IgE responses in vivo, *J. Immunol,* 150, 2717, 1993.

13. **Finkelman, F. d., Madden, K. B., Morris, S. C., Holmes, J. M., Katona, I. M., and Maliszewski, C. R.,** Anticytokine antibodies as carrier proteins: Prolongation of *in vivo* effects of exogenous cytokines by injection of cytokine-anti-cytokine antibody complexes, *J. Immunol.,* 151, 1235, 1993.

14. **Scatchard, G.,** The attractions of proteins for small molecules and ions, *Ann. N.Y. Acad. Sci.,* 51, 660, 1949.

15. **Weiland, G. A., and Molinoff, P. B.,** Quantitative analysis of drug-receptor interactions: I. Determination of kinetic and equilibrium properties, *Life Sciences,* 29, 313, 1981.

16. **Molinoff, P. B., Wolfe, B. B., and Weiland, G. A.,** Quantitative analysis of drug-receptor interactions: III. Determination of the properties of receptor subtypes, *Life Sciences,* 29, 427, 1981.

17. **Lowenthal, J. W., Castle, B. E., Christiansen, J., Schreurs, J., Rennick, D., Arai, N., Hoy, P., Takebe, Y., and Howard, M.,** Expression of high-affinity receptors for murine interleukin-4 (BSF-1) on hemopoietic and nonhemopoietic cells, *J. Immunol.,*140, 456, 1988.

18. **Munson, P. J.,** LIGAND: A computerized analysis of ligand binding data, *Methods Enzymol.,* 92, 543, 1983.

19. **Sharon, M., Klausner, R. D., Cullens, B. R., Chizzonite, R., and Leonard, W. J.,** Novel interleukin-2 receptor subunit detected by cross-linking under high affinity conditions, *Science,* 234, 859, 1986.

20. **Saragovi, H., and Malek, T. R.,** The murine interleukin 2 receptor: Irreversible cross-linking of radiolabeled interleukin 2 to high affinity interleukin 2 receptors reveals a noncovalently associated subunit, *J. Immunol.,* 139, 1918, 1987.

21. **Fernandez-Botran, R., Sanders, V. M., and Vitetta, E. S.,** Interactions between receptors for interleukin 2 and interleukin 4 on lines of helper T cells (HT-2) and B lymphoma cells (BCL$_1$), *J. Exp. Med.,* 169, 379, 1989.

Chapter 8

PREPARATION AND MODIFICATION OF ANTIBODIES

I. POLYCLONAL ANTIBODIES

The choice of animal for antibody production depends mostly upon the amount of antiserum desired. The two most common choices are mice and rabbits. The experimental protocol for successful immunization greatly depends on the type and immunogenicity of an antigen. Therefore, only general points of consideration can be provided here.

The protein antigen is injected intradermally, intramuscularly, or subcutaneously in the presence of adjuvant. At least three different types of adjuvant are commonly used: complete Freund's adjuvant (Sigma), MPL and TDM Emulsion (Ribi Immunochem Research, Inc.), and Gerbu adjuvant (Biotech). The latter two offer significantly higher titers, and, in some cases, even easier handling. On the negative side, their price is significantly higher than that of complete Freund's adjuvant.

Always remember to collect a sample of serum prior to immunization (as a negative control for future tests). Generally, booster immunizations should start 4 to 8 weeks after the first injection and should continue at 3 weeks intervals. For most soluble protein antigens, specific antibodies start to appear in the blood 5 to 8 days after immunization.

The alternative to the use of an adjuvant or a method for increasing the immunogenicity is to couple the protein antigens to a an carrier protein. The two most commonly used carriers are keyhole limpet hemocyanine (KLH) and bovine serum albumin (BSA).

A. PREPARATION OF COMPLETE FREUND'S ADJUVANT-ANTIGEN EMULSION

Materials and Reagents

Complete Freund's adjuvant (CFA; Sigma)
Antigen
3-ml glass syringes
Double-ended locking hub connector
21-G needle

Protocol:

1. Shake the bottle with CFA well to disperse insoluble *Mycobacterium tuberculosis* bacilli.
2. Mix 2 ml CFA with 2 ml of a 2 mg/ml solution of antigen.
3. Draw this mixture up into a 3-ml glass syringe. Attach a double-ended locking hub connector already connected to an empty 3-ml glass syringe. Repeatedly force the mixture back and forth from one syringe to the other. When the mixture is white and quite homogenous, remove the connector and attach a 21-G needle.
4. Remove all air bubbles and inject CFA-antigen emulsion into multiple sites in the animal.

Comments:

1. The volumes shown above could be scaled down according to your individual needs. The use of proportionally smaller syringes will be advantageous.
2. CFA is extremely inflammatory. Always use gloves and protective eyewear.
3. For booster immunization, use incomplete Freund's adjuvant (Sigma) instead of CFA.
4. To determine whether the solution is homogeneous, put a small drop of an antigen-adjuvant mixture onto the surface of 50 ml cold water in a beaker. If the drop disperses, continue mixing the antigen as described in step # 3.

B. RIBI ADJUVANT

This adjuvant consists of purified microbial components in a metabolizable oil. While a combination of monophosphoryl lipid A and synthetic trehalose dicorynomycolate is optimal for the use with strong immunogens, synthetic trehalose dicorynomycolate alone is preferred for regular strength immunogens.

Materials and Reagents

Ribi adjuvant (Ribi Immunochem Research)
Water bath
Syringe
21-G needle
Saline

Antigen
Vortex

Protocol:

1. Warm the vial to 45°C for 10 minutes. Inject 2 ml of saline with desired amount of antigen directly into the vial through the rubber stopper. Vortex vigorously for 3 minutes.
2. Immunize mice with 0.1 ml subcutaneously in two sites or 0.2 ml intraperitonealy. For rats and guinea pigs, inject 0.2 ml subcutaneously in two sites and 0.1 ml intraperitonealy.
3. Repeat on day 21 and test bleed one week later. If necessary, repeat the injection every 21 days.

Comments:

1. Recommended antigen concentration is from 0.05 mg/ml to 0.25 mg/ml in saline.
2. For larger animals (goats, sheep, pigs, and rabbits), use a cell wall skeleton, monophosphoryl lipid A, and synthetic trehalose dicorynomycolate mixture.

II. MONOCLONAL ANTIBODIES

The use of monoclonal antibodies has spread much further than almost any immunological technique. Monoclonal antibodies, for example, are widely used in cell biology, biochemistry, pathology, oncology, hematology, physiology, molecular genetics, and microbiology. The idea of immortalization of specific antibody-forming cells by their fusion with myeloma cells was first introduced by Köhler and Milstein.[1] Since that year, numerous modifications have been published, but all are only variations of the original technique.[2-5]

The idea is basically quite simple. The partner tumor cell line is resistant to the purine analogue 6-thioguanine due to its deficiency of hypoxanthine-guanine phosphoribosyl transferase. The result of this deficiency is a high sensitivity to aminopterin. Normal B lymphocytes are resistant to aminopterin in the presence of hypoxanthine and thymidine (they utilize salvage pathways). While the B cell causes an immortality of the fused cells in HAT medium, the tumor cell causes the immortality in all subsequent feedings.

Materials and Reagents

DME/F12 HYBRI-MAX medium (Sigma) with sodium pyruvate, amino
   acids, antibiotics and 10% FCS (for SP2/0) or 20% FCS
DME/F12 HYBRI-MAX medium without FCS
HAT (Sigma): 1 ml of 50x concentrated powder (diluted in 10 ml of medium
   without FCS) per 50 ml of medium with 20% FCS and 20% CM
HT (Sigma): 1 ml of 50x concentrated powder (diluted in 10 ml of medium
   without FCS) per 50 ml of medium with 20% FCS + 20% CM
Conditioned medium (CM) - i.e., supernatant from SP2/0 cells
NH_4Cl solution, pH 7.2
Polyethylene glycol (PEG) 1,500 (Boehringer-Mannheim, Sigma)
Immunized mice (boosted with antigen 3 days prior to harvesting)
96-well tissue culture plates (Costar)
6-well tissue culture plates (Costar)
24-well tissue culture plates (Costar)
SP2/0-Ag14 cells (ATCC; CRL-1581)
50-ml conical centrifuge tubes
1-ml pipettes
10-ml pipettes
Humidified, 5% CO_2 incubator
Glass beakers

Protocol:

1. Split SP2/0 cells 24 hrs before fusion and cultivate in DME/F12
    HYBRI-MAX medium with 10% FCS.
2. Store conditioned medium.
3. Wash SP2/0 cells three times with serum-free medium by centrifugation
    at 400 x g for 5 minutes at room temperature and count them.
4. Sacrifice mice and store the serum from the mice. It will be used as a
    positive control for later screening.
5. Prepare a spleen cell suspension and place the cells into 50-ml tubes.
    Wash once with serum-free medium by centrifugation at 350 x g for
    10 minutes at room temperature.
6. Lyse the erythrocytes by pouring prewarmed NH_4Cl solution on the
    pellet (approximately 45 ml) and immediately spin. Discard supernatant
    and add 50 ml of serum-free medium. Centrifuge (350 x g for 5
    minutes at room temperature) again. Wash three times with serum-free
    medium under the same conditions. Then resuspend spleen cells in
    serum-free medium and count them.

7. Mix SP2/0 myeloma cells and spleen cells at a 1:2 ratio in a 50-ml tube and fill the tube with serum-free medium. Centrifuge at 500 x g for 5 minutes at room temperature.

8. Prewarm PEG solution to 37°C.

9. Prewarm 20 ml of serum-free medium to 37°C.

10. Discard supernatant from the cell mixture (step # 7)

11. Put the tube with the cell mixture into a 37°C water bath (place a beaker with warm water into the hood). Using a 1-ml pipette, add 1 ml of prewarmed PEG solution, drop-by-drop, stirring after each drop. Resuspend the pellet and let stand for 1 minute.

12. Using new pipettes, add 2 ml of prewarmed serum-free medium drop-by-drop over 2 minutes. Again stir after each drop.

13. Using a 10-ml pipette, add 7 ml of prewarmed medium drop-by-drop over 2 to 3 minutes.

14. Centrifuge the cells at 300 x g for 5 minutes at room temperature.

15. Prewarm medium with 20% FCS and CM to 37°C..

16. Discard the supernatant from step # 14. Put the tube with the cell mixture into a 37°C water bath (place a beaker with warm water into the hood). Break the pellet and count the cells. Dilute the cells to a final concentration of 2.5 x 10^6 cells/ml in complete medium with 20% FCS.

17. Incubate the cells for 2 hours at 37°C.

18. Centrifuge the cells at 300 x g for 10 minutes at room temperature.

19. Resuspend the cells in medium with 20% FCS and CM and adjust the cell concentration 2.5 x 10^6 cells/ml.

20. Put 0.1 ml of the cell suspension into each well of 96-well flat-bottom plates.

21. Incubate overnight at 37°C in a humidified, 5% CO_2 incubator. Add 1 drop of medium with 20% FCS, 20% CM and HAT into each well.

22. Replace the medium after 7 days (but check the plates earlier) with complete medium with HAT.

23. After another 7 days, replace medium with medium with HT.

24. Cells do not require more than one change of complete medium with HT. After step # 23, and subsequently, feed cells using complete DME/F12 HYBRI-MAX medium with sodium pyruvate, amino acids, antibiotics, and 10% FCS. The cells are ready for screening when most of the wells containing growing hybridomas demonstrate 10 to 25% confluence and when those with denser populations turn yellow within two days after feeding.

25. Remove 0.1 ml from each well with growing cells and use it in a screening assay, such as an indirect immunofluorescence or ELISA.

26. Expand the positive wells by transfering the cells into 24-well plates and subsequently into 6-well plates.

27. Repeat screening as described in step # 25.
28. Freeze down all positive cells.
29. Select 10 to 20 best candidate wells and clone them by limiting dilution (see below).

Comments:

1. Do not stir vigorously nor pipette the cell suspension. The fused clumps are very fragile.
2. It is better to put 2 drops using a regular pipette than using a micropipette (step # 20).
3. As a control, the parent (unfused) myeloma SP2/0 cells should be plated in 3 wells at the same concentration as the myeloma cells used in the fusion. Feed following the same schedule as with the fused cells. The control myeloma cells should be dead by day 10. If these cells are not dead, either the parent myeloma cell line has reverted to a non-HAT-sensitive cell line, or the HAT used is not good.
4. Between 1% to 5% of the wells should contain hybridomas secreting the antibody of desired specificity.
5. Several fusion partners are available, SP2/0-Ag14 myeloma cells are probably one of the best choices. It is a nonsecretory myeloma that does not produce either light or heavy chains. Another advantage of this cell line is that these cells also fuse with rat and hamster B lymphocytes.
6. Other good fusion partners are NSO/1 myeloma cells,[6] S194/5XX0 B4.1 myeloma cells[7] and P3-X63-Ag8.653 myeloma cells.[8]
7. The use of BALB/c mice for immunization is recommended because the resulting hybridomas will be entirely of BALB/c origin and, thus, will form an ascites fluid in BALB/c mice.
8. Some authors[9] suggest using 20% horse serum instead of FCS. A gradual switch from horse serum to FCS might be necessary approximately two weeks after fusion.
9. Molecular weights of PEG from 600 to 6,000 are commonly used for hybridization.
10. Some authors[5] recommend to use the fusion solution of 42% PEG and 15% DMSO.
11. The range between 1:1 to 1:10 (myeloma cell:spleen cell) is successful.

A. CLONING BY LIMITING DILUTION

It is quite possible that the selected hybridomas may in fact be formed by two or more clones. Similarly, they might be contaminated by a nonproducing clone. To avoid the risk of future overgrowth by these cells also, cloning is necessary.

166

Materials and Reagents

Hybridoma cells
Humidified, 5% CO_2 incubator
96-well tissue culture plate (Costar)
6-well tissue culture plates (Costar)
24-well tissue culture plates (Costar)
DMEM medium with 10% FCS

Protocol:

1. Harvest the hybridoma cells and count them.
2. Prepare two cell suspensions, 10 ml of each: 50 cells/ml and 5 cells/ml.
3. Fill one half of the wells of a 96-well plate with each dilution (0.2 ml/well).
4. Incubate for 7 to 10 days in a humidified, 5% CO_2 incubator.
5. Check the plates for cell growth. If possible, select clones from wells with a lower dilution. Look for a tight single cluster of cells.
6. Remove 0.1 ml from each well with growing cells and use it in a screening assay, such as an indirect immunofluorescence or ELISA.
7. Expand the positive wells by transfering the cells into 24-well plates and, subsequently, into 6-well plates.

Comments:

1. Another cloning (recloning) might be necessary. Mouse-hamster hybridoma are especially known for an unstable phenotype.

B. GROWING HYBRIDOMAS IN DIALYZING TUBES[10,11]

A well-established method for large scale production of monoclonal antibodies is ascites production in mice. There are, however, several drawbacks of this method. First of all, it is rather expensive. Second, it has become more and more difficult recently to obtain permission to perform these experiments because of restrictions on the use of laboratory animals for this particular purpose. The growing of hybridomas in dialyzing tubes allows the production of monoclonal antibodies in significantly higher quantities than conventional culture in large tissue culture flasks; in addition, such supernatants contain the monoclonal antibodies in a relatively pure form, usually much purer than ascites.

167

Materials and Reagents

Hybridomas
Iscove's medium (Sigma; Gibco) supplemented with 2 mM L-glutamine,
    45 mM sodium pyruvate, antibiotics, 0.05 % of 0.1M mercaptoethanol
    and 10% FCS
FCS
Dialyzing tube, 2.4 cm width, 12,000 to 14,000 molecular weight cutoff
ITS Premix
Primatone RL
Ethanol
Sodium bicarbonate
EDTA-disodium
225-cm² tissue culture flasks (Costar)
Sterile gloves
Roller
Sterile filtered gas mixture

Protocol:

1. Prepare all media.
2. Wash thoroughly all dialyzing tubes with two washings in each of 50% ethanol, 10 mM sodium bicarbonate, 1 mM EDTA, and H_2O.
3. Grow the hybridomas in 225-cm² tissue culture flasks. One day before transfer, change the medium. Spin down the cells by centrifugation at 300 x g for 5 minutes at 4°C, resuspend in complete Iscove's medium and count.
4. Resuspend 1 x 10⁸ cells in 25 ml of Iscove's medium with 2% FCS and 0.05% Primatone RL and 0.1% ITS Premix.
5. Soak dialyzing tubes (30 cm length) in H_2O and make two knots in one end. Sterilize by boiling in H_2O for 2 hours.
6. Place sterile dialyzing tubes into a 225-cm² tissue culture flasks and pour the cell suspension.
7. Seal the dialyzing tubes with two knots.
8. Add 60 ml of Iscove's medium with 2% FCS and 0.05% Primatone RL.
9. Blow a sterile filtered gas mixture (5% CO_2 in air) into the flasks for 30 seconds.
10. Place the flasks on a roller, set the revolutions on 1/5 minutes and incubate at 37°C in a humidified incubator.
11. Change the medium outside the dialyzing tubes and add gas every other day.
12. Harvest supernatant after 10 to 14 days of incubation.

Comments:

1. The washing of dialyzing tubes is necessary to wash away all contaminants from the manufacturing process.
2. From step # 6, perform all handling under sterile conditions.
3. When two people work together, the risk of contamination is greatly reduced.
4. The speed of the roller is very important.

C. PRODUCTION OF ASCITES FLUID

Materials and Reagents

Mice
Pristane (Sigma)
Hybridomas
22-G and 18-G needles
5-ml syringes
RPMI 1640 medium without FCS
50-ml conical centrifuge tubes
15-ml conical centrifuge tubes

Protocol:

1. Using a 22-G needle, inject each mouse intraperitoneally with 0.5 ml of pristane/mouse 7 days prior to injection with hybridoma cells.
2. Harvest hybridoma cells, spin down by centrifugation at 350 x g for 5 minutes at 4°C and count.
3. Resuspend the cells in RPMI 1640 medium without FCS to a concentration of 2.5 x 10^6 cells/ml.
4. Using an 18-G needle, inject each mouse intraperitoneally with 2 ml cells/mouse.
5. Check mice periodically for formation of ascites.
6. Harvest ascites by immobilizing the mouse in one hand and inserting an 18-G needle 1 cm into the abdominal cavity. Allow the fluid to drip into 15-ml conical centrifuge tubes.
7. Mix the ascites with a wooden stick before centrifugation. Centrifuge the ascites at 1,500 x g for 10 minutes at 4°C. Harvest supernatant and store at -20°C.

169

Comments:

1. Use BALB/c mice (or other syngeneic strain in the case that other than usual BALB/c mice were used for original immunization). If previous attempts to produce ascites fluid failed, use more expensive athymic nude mice.
2. The total amount of cells/mouse may vary between individual hybridomas. If the mice form ascites fluid faster than 5 to 10 days after injection, decrease the number of cells in a future experiment. If no ascites fluid is formed after 3 weeks, repeat inoculation with an increased number of cells.
3. Alternatively, use mice irradiated by a sublethal dose of irradiation.
4. The ascites fluid might be under high pressure, so be sure that the hub of the needle is pointed into a tube before inserting into the peritoneum.
5. Reposition or reinsertion of the needle might be necessary during harvesting of ascites fluid.
6. Wait for a full development of ascites; do not harvest the first drops of fluid.

III. COUPLING OF SYNTHETIC PEPTIDES TO CARRIER PROTEINS

Most of the synthetic peptides used for antibody production are too small to be immunogenic. Generally, peptides with a length of 10 to 15 residues are used to make sera that react with the native protein. It is highly recommended to use the C-terminal sequence if the peptide is hydrophilic or the N-terminal sequence if it is hydrophobic.[12] A selected carrier protein should be a good immunogen and have a sufficient number of amino acid residues with reactive side chains. The principal carriers used most for coupling peptides include KLH (MW over 2,000 kDa), bovine serum albumin (MW 67 kDa), tetanus toxoid (MW 150 kDa), ovalbumin (MW 43 kDa), and myoglobin (MW 17 kDa). Bovine serum albumin has the disadvantage that anti-carrier antibodies present in raised serum might be a problem if the serum is used in the presence of fetal calf serum.

The most commonly used reagents for peptide protein coupling are glutaraldehyde, 1-ethyl-3-(3-dimethylaminopropyl)-carbodiimide, *bis*-imido-esters, *bis*-diazotized benzidine and m-maleimidobenzoyl-N-hydroxysuccimide ester.[13] The choice of the coupling reagent will be based on the nature of free groups. Peptides corresponding to the amino terminus of protein should be coupled via their carboxy-terminal residue, whereas peptides corresponding to the carboxyl terminus should be coupled via their amino-terminal residue.

170

A. COUPLING OF SYNTHETIC PEPTIDE TO CARRIER PROTEIN USING GLUTARALDEHYDE[14]

Glutaraldehyde cross-links the carrier and peptide via their amino groups. Therefore, peptides having Lys residues at positions other than the amino terminus should not be coupled using this technique. The best KLH can be obtained from Pacific Bio-Marine Laboratories.

Materials and Reagents

Borate buffers, pH 10.0 and pH 8.5
Synthetic peptide
Keyhole limpet hemocyanin (KLH; Calbiochem; Sigma; Pacific Bio-Marine Laboratories)
0.3% glutaraldehyde solution in borate buffer, pH 10.0
1 M glycine
15-ml glass tube
Stir plate
Dialyzing tube

Protocol:

1. Dissolve 10 mg of KLH in 2 ml of borate buffer, pH 10.0, in a 15-ml glass tube. While mixing on a stir plate, add 10 mg of synthetic peptide.
2. Slowly add 1 ml of 0.3% glutaraldehyde solution (freshly made) with constant mixing on a stir plate. Incubate for 2 hours at room temperature.
3. Add 0.25 ml of 1 M glycine to block unreacted glutaraldehyde. Incubate for 30 minutes at room temperature.
4. Dialyze against borate buffer, pH 8.5, overnight at 4°C. Replace with fresh borate buffer and dialyze again overnight at 4°C.

B. COUPLING OF SYNTHETIC PEPTIDE TO CARRIER PROTEIN USING M-MALEIMIDOBENZOYL-N-HYDROXYSUCCINIMIDE ESTER[14]

Materials and Reagents

m-maleimidobenzoyl-N-hydroxysuccinimide ester (MBS; Pierce)
Keyhole limpet hemocyanin (KLH; Calbiochem; Sigma; Pacific Bio-Marine Laboratories)

171

0.01 M phosphate buffer, pH 7.0
Dimethylformamide (Sigma)
0.05 M phosphate buffer, pH 6.0
Cys-containing synthetic peptide
PBS
0.1 M HCl
0.1 M NaOH
Dialysis tube (10,000 molecular weight cutoff)
3-ml and 15-ml glass tubes
PD-10 column (Pharmacia)
Stir plate
Fraction collector
Spectrophotometer

Protocol:

1. Dissolve 5 mg KLH in 0.5 ml of 0.01 M phosphate buffer, pH 7.0.
2. Dialyze against 0.01 M phosphate buffer, pH 7.0, overnight at 4°C. Put the dialyzed solution into a 3-ml glas tube.
3. Add 70 µl of 15 mg/ml MBS in dimethylformamide (prepared fresh) to the dialyzed solution and gently mix on a stir plate for 30 minutes at room temperature.
4. Prepare a PD-column by washing with 50 ml of 0.05 M phosphate buffer, pH 6.0. Load the mixture from step # 3 and elute with 0.05 M phosphate buffer, pH 6.0. Collect 0.5 ml fractions and check the A_{280} of these fractions. The first peak represents maleimide-activated KLH, the second peak represents free MBS.
5. Pool the maleimide-activated KLH fractions in a 15-ml glass tube.
6. Dissolve 5 mg synthetic peptide in 1 ml PBS.
7. Mix the pooled fraction with the peptide solution. Adjust pH to 7.3 and incubate for 3 hours at room temperature with constant mixing.
8. Dialyze against water overnight at 4°C. Replace with fresh water and dialyze overnight again at 4°C.

Comments:

1. If the peptide is not soluble in PBS, try 6 M guanidine-HCl/0.01 M phosphate buffer, pH 7.0.

C. COUPLING OF SYNTHETIC PEPTIDE TO CARRIER PROTEIN USING 1-ETHYL-3-(3-DIMETHYLAMINOPROPYL)-CARBODIIMIDE

(3-Dimethylaminopropyl)-carbodiimide[15,16] cross-links the free amino acid group of a carrier with a C-terminal carboxyl group of the peptide. Do not use peptides with internal Asp, Glu, or Lys residues.

Materials and Reagents

1-Ethyl-3-(3-dimethylaminopropyl)-carbodiimide-HCl (EDCI; Sigma)
5-ml glass tube
Synthetic peptide
Keyhole limpet hemocyanin (KLH; Calbiochem; Sigma; Pacific Bio-Marine Laboratories)
Stir plate
Dialysis tube (10,000 molecular weight cutoff)

Protocol:

1. Dissolve 10 mg peptide in 1 ml water in a 5-ml glass tube.
2. Add 40 mg EDCI to this solution with constant gentle stirring. Adjust pH to 4.5 and incubate for 10 minutes at room temperature.
3. Add 0.5 ml KLH (5 mg/ml in water) to this solution. Incubate for 2 hours at room temperature.
4. Dialyze against water overnight at 4°C. Replace with fresh water and dialyze again overnight at 4°C.

D. COUPLING OF SYNTHETIC PEPTIDE TO CARRIER PROTEIN USING *BIS*-DIAZOTIZED BENZIDINE [14]

Bis-diazotized benzidine cross-links Tyr residues in the carrier protein to Tyr residues in peptides.

Materials and Reagents

Benzidine-HCl (Sigma)
0.2 M HCl
NaNO$_2$
Borate buffer, pH 9.0
15-ml test tube

173

Beaker
Keyhole limpet hemocyanin (KLH; Calbiochem; Sigma; Pacific Bio-Marine
　　Laboratories)
Synthetic peptide
Dialyzing tube
Stir plate
Ice

Protocol:

1. Dissolve 5 mg benzidine in 1 ml of 0.2 M HCl. Add 3.5 mg NaNO$_2$ and incubate 1 hour with constant stirring at 4°C in the dark.
2. Dissolve 5 mg KLH in 10 ml borate buffer, pH 9.0, in a beaker. Add 2 mg synthetic peptide and cool on ice.
3. Add dropwise a *bis*-diazotized benzidine solution from step # 1 to the solution from step # 2. Adjust pH to 9.0 and incubate with constant stirring for 2 hours at 4°C.
4. Dialyze against water overnight at 4°C. Replace with fresh water and dialyze overnight again at 4°C.

Comments:

1. Benzidine and *bis*-diazotized benzidine are carcinogens.

REFERENCES

1. **Kohler, G. and Milstein, C.,** Continuous cultures of fused cells secreting antibody of predefined specificity, *Nature,* 256, 495, 1975.
2. **Springer, T. A.,** Ed., in *Hybridoma Technology in the Biosciences and Medicine,* Plenum Press, New York, 1985.
3. **Goding, J. W.,** *Monoclonal Antibodies: Principles and Practise,* Academic Press, San Diego, 1986.
4. **Oi, V. T. and Herzenberg, L. A.,** Immunoglobulin-producing hybrid cell lines, in *Selected Methods in Cellular Immunology,* Mishell, B. B. and Shiigi, S. M., Eds., W. H. Freeman, New York, 1980, 351.
5. **Zola, H.,** *Monoclonal Antibodies: A Manual of Techniques,* CRC Press, Boca Raton, 1985.
6. **Galfre, G. and Milstein, C.,** Preparation of monoclonal antibodies: Strategies and procedures, *Meth. Enzymol.,* 73, 3, 1981.

7. **Trowbridge, I. S.**, Interspecies spleen myeloma hybrid producing monoclonal antibodies against mouse lymphocyte surface glycoprotein T200, *J. Exp. Med.*, 148, 313, 1978.

8. **Kearney, J. F., Radbruch, A., Liesegang, B. and Rajewsky, K.**, A new mouse myeloma cell line that has lost immunoglobulin expression but permits the construction of antibody-secreting hybrid cell lines, *J. Immunol.*, 123, 1548, 1979.

9. **Eshhar, Z.**, Monoclonal antibody strategy and techniques, in *Hybridoma Technology in the Biosciences and Medicine*, Springer, T.A., Ed., Plenum Press, New York, 1985, 3.

10. **Sjogren-Jansson, E. and Jeansson, S.**, Growing hybridomas in dialysis tubing: optimalization of the technique, in *Laboratory Methods in Immunology, Vol.I*, Zola, H., Ed., CRC Press, Boca Raton, 1990, 41.

11. **Sjogren-Jansson, E. and Jeansson, S.**, Large-scale production of monoclonal antibodies in dialysis tubing, *J. Immunol. Methods*, 84, 359, 1985.

12. **Maloy, W. L. and Coligan, J. E.** Selection of immunogenic peptides for antisera production, in *Current Protocols in Immunology*, Coligan, J. E., Kruisbeek, A. M., Margulies, D. H., Shevach, E. M. and Strober, W., Eds., Green Publishing and Wiley Interscience, New York, 1994, 9.3.1.

13. **Van Regenmortel, M. H. V., Briand, J. P., Muller, S. and Plaue, S.**, Synthetic polypeptides as antigens, in *Laboratory Techniques in Biochemistry and Molecular Biology, Vol. 19*, Burdon, R. H. and von Knippenberg, P. H., Eds., Elsevier, Amsterdam, 1988, 361.

14. **Maloy, W. L., Coligan, J. E. and Paterson, Y.**, Production of antipeptide antisera, in *Current Protocols in Immunology*, Coligan, J. E., Kruisbeek, A. M., Margulies, D. H., Shevach, E. M. and Strober, W., Eds., Green Publishing and Wiley Interscience, New York, 1991, 9.4.1.

Chapter 9

DETECTION OF ANTIBODIES

I. HEMOLYTIC PLAQUE ASSAYS

The hemolytic plaque assays were originally developed to visualize the small amount of lytic antibody released into the vicinity of a single antibody-forming lymphocyte.[1-3] The idea is simple: secreted anti-erythrocyte antibodies bind to the surrounding red blood cells, after addition of complement the indicator erythrocytes lyse. The antibody-producing cell is visualized by formation of an easily visible plaque. The original method (so called direct method) measured only formation of IgM antibodies. The reason is that IgM antibodies fix complement efficiently, whereas IgG and IgA antibodies do not. The formation and secretion of antibodies of other isotypes can be measured using an indirect method employing anti-Ig sera. These antibodies bind to the specific anti-erythrocyte antibodies already bound to the surface of erythrocytes. In order to distinguish between the number of IgM-formed plaques and plaques formed by other isotypes of immunoglobulins, it is possible to either subtract the number of IgM plaques from a simultaneous plate or block the formation of IgM plaques before the indirect assay.

By using Protein A-coupled red blood cells[4,5] you can evaluate the total number of lymphocytes producing immunoglobulins of all specificities. Another modification of this technique is either to attach various haptens to indicator erythrocytes and subsequently measure antibodies of anti-hapten specificity or measure the polyclonal formation of anti-erythrocyte antibodies after immunization with different antigens.[6,7]

Several substantial modification of the plaque-forming assay has been developed. The most important versions are the original Jerne and Nordin's[1] using Petri dishes, Sterzl and Mandel's[2] using drop technique and Mishell and Dutton's[8] modification using microscopic slides. All three versions will be explained with direct IgM methods and only one of these techniques will be shown with indirect or Protein A methods.

A. DIRECT METHOD USING PETRI DISHES

Materials and Reagents

Bacto agar (Difco Laboratories)
100 x 15-mm Petri dishes
Water bath
Incubator at 37°C
Sheep erythrocytes (SRBC; Cleveland Scientific)
Cell suspension
DEAE dextran, MW 500 to 2,000 kDa (Pharmacia Fine Chemicals)
Guinea pig complement
Pipettes
Glucose-phosphate buffered saline (G-PBS)
2X and 1X balanced salt solution (BSS)
5-ml tubes

Protocol:

1. Dissolve 0.7 g Difco agar in 50 ml of distilled water by boiling. Prewarm 50 ml of 2X BSS containing 0.1 g of DEAE dextran at 45°C and add agar solution.
2. Prewarm tubes in water bath to 45°C.
3. Add 2 ml of the agar mixture into each tube.
4. Wash SRBC at least three times in G-PBS by centrifugation at 1,000 x g and resuspend to 15% (v/v) in G-PBS.
5. Dilute the tested cell suspension in G-PBS.
6. Add 0.1 ml of SRBC to 10 tubes in the water bath. Add cell suspension to one tube and mix the tube. Pour rapidly over the Petri dish, tilting the dish so the entire surface is covered. Cover the dish (but leave it slightly ajar) and let solidify for a few minutes before moving. After all dishes are poured, transfer them into an incubator and incubate at 37°C for 1 hour. If the incubator is not humidified, add a large container of water.
7. Add 1.5 ml of 1:10 diluted (in 1X BSS) guinea pig complement to each dish. Dilute the complement immediately before use and keep it on ice.
8. Incubate at 37°C for 1 hour.
9. Count the plaques against the light. Questionable plaques can be verified by microscopic examination. The true plaque will have a lymphocyte in the center.

Comments:

1. The optimal dilution of the lymphocyte suspension must be determined by pilot experiments. The ideal dilution is a dilution that contains 100 to 300 plaques per Petri dish. If the level of immune response is unknown, prepare several dilutions at 10-fold increments. It is important to keep the volume low in order not to dilute the indicator SRBC.
2. Preabsorb the guinea pig complement with SRBC prior use. Aliquot the absorbed complement into small volumes and store at -80°C.
3. Keep the cell suspension on ice so the cells will not release any antibodies before their transfer to the test tubes.

B. INDIRECT METHOD USING PETRI DISHES

Materials and Reagents

Bacto agar (Difco Laboratories)
100 x 15-mm Petri dishes
Water bath
Incubator at 37°C
Sheep erythrocytes (SRBC; Cleveland Scientific)
Cell suspension
DEAE dextran, MW 500 to 2,000 kDa (Pharmacia Fine Chemicals)
Guinea pig complement
Pipettes
Glucose-phosphate buffered saline (G-PBS)
2X and 1X balanced salt solution (BSS)
5-ml tubes
Anti-Ig serum

Protocol:

1. Prepare one set of Petri dishes for the direct IgM method as described above. In parallel, prepare the second identical set of Petri dishes for the indirect IgG method.
2. After 1 hour of incubation at 37°C, add 1.5 ml of BSS to the direct dishes and 1.5 ml of anti-Ig serum diluted in BSS.
3. Incubate for 1 hour at 37°C, pour of the BSS and anti-Ig serum and add 1.5 ml of diluted guined pig complement. Incubate and count as described above.

Comments:

1. Pretest the optimal dilution of anti-Ig serum, aliquot into small volume and store at -30°C.
2. Prepare at least two different concentration of cells for the indirect set. Generally, the IgG response is much higher than primary, thus the number of cells should be adequately lower.

C. DIRECT METHOD USING SLIDES

Materials and Reagents

Microscope slides (CMS)
SRBC
Bacto agar (Difco Laboratories)
Water bath
Incubator at 37°C
Cell suspension
Guinea pig complement
Pipettes
Agarose (#162-0100, Bio-Rad Laboratories; or SeaPlaque from FMC Bioproducts)
12 x 75-mm glass tubes (CMS)
15-ml centrifuge tubes
RPMI 1640 medium
95% ethanol
Slide assay jar
Brush (2 to 4 in. wide)

Protocol:

1. Clean the slide by submersion into the slide jar filled with 95% ethanol for 5 minutes. Allow slides to dry by evaporation.
2. Prepare a 0.1% solution (w/v) of agarose in distilled water by boiling. Coat the clear portion of each slide with hot agarose using a brush. Do not cover the frosted ends of the slide. Allow the slide to solidify and store them in covered container. If stored at 4°C, the slides can be stored indefinitely.
3. Prepare 0.5% agarose in RPMI 1640 medium and keep it in water bath at 45°C.

4. Prewarm the tubes in the water bath.
5. Wash SRBC three times by centrifugation at 1,000 x g in RMPI 1640 medium and resuspend them to 6.6% (v/v) in medium.
6. Wash cell suspension in cold medium and keep it on ice.
7. Pour 1 ml of 0.5 % agarose into each tube, add 100 µl of SRBC and add cell suspension into one tube. Mix the tube and pour its contents on two slides. Allow the slides to solidify for a few minutes.
8. Add 1 to 2 ml of diluted complement and incubate for 3 hours at 37°C.
9. Count the plaques against the light. Questionable plaques can be verified by microscopic examination. The true plaque will have a lymphocyte in the center.

Comments:

1. The main advantage of this modification is that it uses less reagents. Microscope slides are also cheaper than Petri dishes.

D. DROP PLAQUE TECHNIQUE[2]

Materials and Reagents

100 x 15-mm Petri dishes
Water bath at 45°C
RPMI 1640 medium
PBS
FCS
Cell suspenison
Bacto agar (Difco Laboratories)
12 x 75-mm glass tubes (CMS)
DEAE dextran, MW 500 to 2,000 kDa (Pharmacia Fine Chemicals)
Guinea pig complement
Glass pipettes
Gas burner

Protocol:

1. Dissolve agar (12.5 mg/ml) together with DEAE dextran (0.75 mg/ml) in distilled water.
2. Wash cell suspension in ice cold RPMI 1640 (pH 7.3 to 7.4) by centrifugation at 250 x g and resuspend it in the same medium. Count and resuspend the cell suspension at optimal concentration, not exceeding 5 x 10^7 cells/ml.

181

3. Put 0.3 ml of the cell suspension into glass tubes and keep them on ice.
4. Wash SRBC at least three times in PBS by centrifugation at 1,000 x g and resuspend to 5% (v/v) in PBS. Put the SRBC suspension in the water bath (45°C).
5. Prewarm the tubes in the water bath (45°C). Add 4 ml of warm agar into each tube.
6. At the same time put 0.5 ml 5% FCS and 0.5 ml RPMI 1640 into the tube and incubate in the water bath at 45°C for 5 minutes.
7. Mix the agar solution and the RPMI 1640 medium with 5% FCS. Add 1.7 ml of prewarmed SRBC.
8. Prewarm 0.3 ml of cell suspension (previously kept on ice) for approximately one minute in the water bath and add 0.6 ml of agar - SRBC suspension using prewarmed (one passage through a flame) glass pipettete.
9. Mix the tube shortly and drop the final suspension using the same prewarmed pipette—from approximately a 50-cm height on a Petri dish (approx. 8 to 10 individual drops). Rotate the dish.
10. Allow the Petri dishes to solidify for a few minutes before moving. After all dishes are poured, transfer them into an incubator and incubate at 37°C for 2 hours. If the incubator is not humidified, add a large container of water.
11. Add 1.5 ml of 1:10 diluted (in 1X BSS) guinea pig complement to each dish. Dilute the complement immediately before use and keep it on ice.
12. Incubate at 37°C for 1 hour.
13. Count the plaques against the light. Questionable plaques can be verified by microscopic examination. The true plaque will have a lymphocyte in the center.

E. ANTI-HAPTEN METHOD

Any of the plaque techniques described above can be used for detection of anti-hapten antibody formation. The only difference is the preparation of hapten-coated sheep red blood cells. We describe the two most commonly used haptens, trinitrophenyl (TNP) and azophenyl arsonate (Ars), but numerous other haptens including azophenyl lactoside, azophenyl glucoside, dinitrophenyl and fluorescein isothiocyanate can be used.[7]

1. Coupling of TNP to SRBC

__Materials and Reagents__

SRBC

FCS
G-PBS
Picrylsulfonic acid, (2,4,6-trinitrobenzene sulfonic acid; TNBS; Sigma)
0.28 M cacodylate buffer, pH 6.9
15-ml conical centrifuge tubes
Glass beaker
Aluminium foil

Protocol:

1. Wash SRBC at least three times in G-PBS by centrifugation at 1,000 x g at 4°C.
2. Dissolve 20 mg of 2,4,6-trinitrobenzene sulfonic acid in 7 ml of cacodylate buffer.
3. Add 1 ml of packed SRBC (100% suspension) very slowly to the 2,4,6-trinitrobenzene sulphonic acid solution and mix for 10 minutes with a magnetic stirring bar.
4. Wash SRBC in cold G-PBS with 1% FCS by centrifugation at 400 x g at 4°C until the supernatant is colorless.
5. Resuspend the SRBC to optimal dilution in buffer according to the type of plaque assay.

Comments:

1. Steps # 2 to 4 have to be done in aluminium foil - covered glassware in order to protect against light.

2. Coupling of SRBC with Ars

Materials and Reagents

SRBC
Arsanilic acid (Sigma)
1 M HCl
NaNO$_2$ (15 mg/ml in H$_2$O)
FCS
G-PBS
1 M phosphate buffer, pH 7.6
15-ml conical centrifuge tubes
Glass beaker

Protocol:

1. Dissolve 70 mg of arsanilic acid in 1.2 ml of ice-cold 1 M HCl.
2. Add 1.6 ml of ice-cold $NaNO_2$ solution, mix and incubate for 10 minutes on ice.
3. Add 34 ml of ice-cold H_2O and incubate for additional 40 minutes on ice.
4. Add another 34 ml of ice-cold H_2O, divide into 2 ml aliquots, freeze at -80°C and store.
5. Wash SRBC at least three times in G-PBS by centrifugation at 1,000 x g at room temperature and resuspend to 50% (v/v) in G-PBS.
6. Mix 2 ml of diazonium phenyl arsonate (prepared in steps 2 to 4) with 1 ml of 1 M phosphate buffer and add this mixture to 2 ml of 50% SRBC.
7. Incubate at room temperature for 20 minutes with occasional shaking.
8. Wash SRBC in cold G-PBS with 1% FCS by centrifugation at 400 x g at 4°C at least three times.
9. Resuspend the SRBC to optimal dilution in buffer according to the type of plaque assay.

3. Total Production of Immunoglobulins

Materials and Reagents

SRBC
Protein A, 0.5 mg/ml in 0.85% NaCl (Pharmacia Fine Chemicals; Sigma)
0.85% (w/v) NaCl
$CrCl·6 H_2O$, 6.6 mg/ml in 0.85% NaCl (Sigma)
Rotator
Rabbit anti-mouse Ig (or class specific antibody)
15-ml conical centrifuge tubes with cap

Protocol:

1. Wash SRBC at least three times in 0.85% NaCl by centrifugation at 1,000 x g at room temperature.
2. Mix 1 ml of packed SRBC with 1 ml of Protein A solution and 10 ml of $CrCl_3$ solution diluted 1:100 in 0.85% NaCl.
3. Incubate with constant slow rotation for 45 minutes at 37°C.
4. Wash SRBC in 0.85% NaCl centrifugation at 400 x g at 4°C at least three times. Resuspend the SRBC to optimal dilution in buffer according to the type of plaque assay.

Comments:

1. This method is useful either when the information about total Ig synthesis will supplement study of IgM and/or IgG production or in experiments where the lymphocytes were polyclonally activated (such as after injection with LPS or after PWM addition *in vitro*).
2. Some authors[6] recommend using a $CrCl_3$ solution prepared immediately before the coupling of Protein A to the erythrocytes. However, we have found that this solution can be stored at room temperature for at least two years and that optimal results are achieved using a solution that has been stored for at least three months.

II. ELISA

Enzyme-linked immunosorbent assay (ELISA)[9] is a technique for assaying the presence of antibodies in various fluids. Using any of numerous modifications, this method allows us to qualitatively or quantitatively evaluate a particular antibody activity, to measure an antigen using a defined antibody preparation, or to detect cell-surface antigens. ELISA is one of the most versatile and widely spread techniques, and its sensitivity is between 100 pg/ml and 1 ng/ml.

As a general rule, the incubation times used in individual steps might be either overnight at 4°C or 2 hours at either room temperature or 37°C. The term *incubate* will be used instead of repeating *overnight at 4°C or 2 hours at either room temperature or 37°C* in every step.

It is very difficult to find two laboratories performing this assay exactly the same way. The readers should try to use the following information as a basis for development of their own modification. One of the more important modifications is the use of 0.25% gelatin or 5% instant milk instead of bovine serum albumin.

Various substrates used for visualization of an ELISA assays are shown in Table 1.

A. INDIRECT ELISA

This technique is used primarily for screening antibodies such as in testing of hybridoma culture supernatants. It does not require the use of preexisting specific antibodies. The wells are coated with the antigen followed by incubation with supernatants. Unbound antibodies are washed out and an Ig-specific antibody conjugated to an enzyme is used for visualization.

185

Materials and Reagents

ELISA reader
Antigen
Test samples
Enzyme-conjugated antibody
96-well U- or flat-bottom microtiter plates (Immulon; Dynatech)
Substrate
PBS containing 0.05% Tween 20 (PBS-Tween)
PBS containing 0.05% Tween 20, 1% BSA and 0.02% NaN_3 (PBS-BSA-Tween)
Multichannel pipette
Carbonate-bicarbonate buffer, pH 9.6
Plastic squirt bottle
Paper towels

Protocol:

1. Dilute the antigen in carbonate-bicarbonate buffer at a concentration of 5 µg/ml. Using a multichannel pipette, dispense 100 µl of antigen solution into each well on each plate. This coats the plates with antigen.
2. Cover the plates with a cover or wrap them in plastic wrap and incubate them.
3. Rinse the plates five times over a sink by filling all wells with PBS-Tween from a plastic squirt bottle. Flick the PBS-Tween into the sink after each rinse.
4. Block residual binding capacity of wells by filling all wells with PBS-BSA-Tween and incubating for 30 minutes at room temperature.
5. Rinse the plates five times over a sink by filling all wells with PBS-Tween from a plastic squirt bottle. Flick the PBS-Tween into the sink after each rinse. After the last wash, remove residual PBS-Tween by laying the plates face down on paper towels for several seconds.
6. Add 100 µl of test samples diluted in PBS-BSA-Tween to each well. Cover the plates and incubate.
7. Rinse the plates five times over a sink by filling all wells with PBS-Tween from a plastic squirt bottle. Flick the PBS-Tween into the sink after each rinse. After the last wash, remove residual PBS-Tween by laying the plates face down on paper towels for several seconds.

TABLE 1

Various Substrates Used for Visualization of an ELISA

Enzyme	Substrate	OD
Alkaline phosphatase[a]	p-nitrophenyl phosphate	405
	4-methylumbelliferyl phosphate	365/450
Horseradish peroxidase	O-phenylendiamine	492
	2,2'-azino-bisi(3-ethylbenzthiazoline sulfonic acid	414
	O-dianisidine	530
	5-aminosalicylic acid	474
	3,3',5,5'-tetramethylbenzidine	450
β-D-galactosidase	O-nitrophenyl-β-D-galactopyranoside	420
	chlorophenolic red-β-D-galactopyranoside	574
	resorufin-β-D-galactopyranoside	570
Urease	bromcresol purple	588
Acetylcholine esterase	Ellman's reagent	412

[a]The fluorogenic system using 4-methylumbelliferyl phosphate is up to 100 times faster than using p-nitrophenyl phosphate.

Note: Readers seeking more information about various enzyme - substrate combinations should see reference 10.

8. Add 100 µl of enzyme-conjugated antibody diluted in PBS-BSA-Tween to each well. Cover the plates and incubate.

9. Rinse the plates three times over a sink by filling all wells with PBS-Tween from a plastic squirt bottle. Flick the PBS-Tween into the sink after each rinse. After the last wash, remove residual PBS-Tween by laying the plates face down on paper towels for several seconds.

10. Add 100 µl of substrate solution to each well, cover the plates and incubate them for 1 hour at room temperature in the dark. Check the color development occasionally. Read the reaction on an ELISA reader using the appropriate filter.

Comments:

1. The optimal concentration of enzyme-conjugated antibody should be determined by a pilot experiment utilizing criss-cross serial dilution analysis.
2. Covered antigen-coated plates can be stored at 4°C for several months. Do not let them dry out.
3. The optimal incubation time in the last step depends on the type of enzyme-substrate combination and on the concentration of antibody in tested samples. Stop the reaction (either by reading or by adding a stopping reagent) when positive wells reach the desired intensity.

B. ANTIBODY - SANDWICH ELISA FOR DETECTION OF SOLUBLE ANTIGENS

The wells are coated with a specific antibody followed by incubation with antigen-containing solutions. Unbound material is washed out and different enzyme-conjugated antigen-specific antibodies are used for visualization.

Materials and Reagents

ELISA reader
Specific antibody
Tested solution
Enzyme-conjugated antibody
96-well U- or flat-bottom microtiter plates (Immulon; Dynatech)
Substrate
PBS containing 0.05% Tween 20 (PBS-Tween)
PBS containing 0.05% Tween 20, 1% BSA and 0.02% NaN$_3$ (PBS-BSA-Tween)
Multichannel pipette
Carbonate-bicarbonate buffer pH 9.6
Plastic squirt bottle
Paper towels

Protocol:

1. Dilute the specific antibody in carbonate-bicarbonate buffer at a concentration of 5 µg/ml. Using a multichannel pipette, dispense 100 µl of antibody solution into each well on each plate. This coats the plates with the specific antibody.

188

2. Cover the plates with a cover or wrap them in plastic wrap and incubate them.
3. Rinse the plates five times over a sink by filling all wells with PBS-Tween from a plastic squirt bottle. Flick the PBS-Tween into the sink after each rinse.
4. Block residual binding capacity of wells by filling all wells with PBS-BSA-Tween and incubating for 30 minutes at room temperature.
5. Rinse the plates five times over a sink by filling all wells with PBS-Tween from a plastic squirt bottle. Flick the PBS-Tween into the sink after each rinse. After the last wash, remove residual PBS-Tween by laying the plates face down on paper towels for several seconds.
6. Dilute antigen (both standard antigen as a positive control and a tested antigen) in PBS-BSA-Tween (optimal range is from 0.1 to 1000 ng/ml).
7. Add 100 µl of diluted antigen into antibody-coated wells and incubate.
8. Rinse the plates five times as described in step # 5.
9. Add 100 µl of test samples diluted in PBS-BSA-Tween to each well. Cover the plates and incubate.
10. Rinse the plates five times as described in step # 5.
11. Add 100 µl of enzyme-conjugated antibody diluted in PBS-BSA-Tween to each well. Cover the plates and incubate.
12. Rinse the plates five times as described in step # 5.
13. Add 100 µl of substrate solution to each well, cover the plates and incubate them for 1 hour at room temperature in the dark. Check the visualization occasionally. Read the reaction on ELISA reader using the appropriate filter.
14. Prepare a standard curve from data obtained from antigen standards.
15. Interpolate the concentration of antigen in the test samples from that standard curve.

Comments:

1. Optimal concentration of specific antibody and enzyme-conjugated antibody should be determined in a pilot experiment by criss-cross serial dilution analysis.
2. Covered antibody-coated plates can be stored at 4°C for several months. Do not let them dry out.
3. The optimal incubation time in the last step depends on the type of enzyme-substrate combination and on the concentration of antibody in tested samples. Stop the reaction (either by reading or by adding of stopping reagent) when positive wells reach desired intensity.

C. DIRECT CELLULAR ELISA[11]

This technique is used for quantitative evaluation of cell-surface determinants using existing enzyme-conjugated antibodies specific for that particular surface marker. Unbound antibody is washed away and a substrate is added for visualization.

Materials and Reagents

ELISA reader
Enzyme-conjugated antibody
96-well U-bottom microtiter plates (Immulon; Dynatech)
Substrate
Cells
PBS containing 0.05% Tween 20 (PBS-Tween)
Multichannel pipette
Carbonate-bicarbonate buffer pH 9.6
Plastic squirt bottle
Paper towels
15-ml conical centrifuge tubes
RPMI 1640 medium with 5% FCS
Vortex

Protocol:

1. Prepare cell suspension in RPMI 1640 medium containing 5% FCS and adjust cell density to 2.5×10^5 cells/ml.
2. Dispense 100 µl of the diluted cell suspension into each well. Centrifuge the plate at 450 x g for 2 minutes at 4°C. Decant the supernatant and gently mix the plate using a vortex to disrupt pellet.
3. Add 100 µl of the enzyme-antibody conjugate diluted in PBS-Tween into each well and incubate for 2 hours at 4°C.
4. Centrifuge the plate at 450 x g for 2 minutes at 4°C. Decant the supernatant and gently mix the plate to disrupt pellet. Resuspend the cells in 200 µl of PBS-Tween. Repeat twice.
5. Add 100 µl of substrate and incubate until the reaction has reached the desired levels at room temperature in the dark.
6. Read the reaction on an ELISA reader using the appropriate filter.

Comments:

1. The method is suitable for cell lines or homogenous populations of cells.
2. You can easily convert this method into an indirect ELISA by either substituting unlabeled antibody for the enzyme-conjugated antibody, followed by a second incubation with Ig-specific enzyme-conjugated antibody, or by substituting biotinylated antibody for the enzyme-conjugated antibody, followed by a second incubation with an avidin-enzyme conjugate.
3. The optimal concentration of cells should be determined in a pilot experiment. The most common concentration is in the range of 1 to 5 x 10^5 cells/ml.
4. The optimal incubation time in the last step depends on the type of enzyme-substrate combination and on the concentration of antibody in the test samples. Stop the reaction (either by reading or by adding of stopping reagent) when positive wells reach desired intensity.
5. Some types of cells express substantial levels of alkaline phosphatase. A pilot experiment measuring the spontaneous level of alkaline phosphatase (cells and substrate only) must be performed. If these levels are too high, another enzyme-substrate combination must be used.
6. Do not fix the cells unless it is known that the particular surface determinant retains its antigenicity after fixation. In that case, use a 0.5% glutaraldehyde solution in PBS and incubate for 30 minutes at room temperature
7. It is also possible to incubate the cells in wells for 24 hours at 37°C (step # 1). In that case, incubate under sterile conditions.

D. ELISPOT ASSAY[12-15]

The antibodies released from specific antibody-secreting cells are immobilized at the point of release via attachment to an underlying solid phase to which antigens have been chemically conjugated. The captured antibody is then subsequently visualized by the application of enzyme-conjugated antibody and an enzyme substrate. Cell products secreted by a single cell generates spots that are permanent. The colored end product is very stable and can be enumerated later. This method is more sensitive than conventional plaque techniques and overcomes their possible disadvantages such as batch-to-batch variations in the susceptibility of target erythrocytes. The recent modifications have improved the sensitivity of the ELISPOT technique such that cells producing as few as 100 molecules of the test protein per second can be detected. The ELISPOT technique can be used for detection of cells secreting antibodies against any antigen, soluble or particulate, that can be absorbed onto a solid surface. This method can be easily modified for enumeration of cytokine-secreting cells[14] by substitution of an anti-Ig antibody with an anti-cytokine antibody.[16]

Materials and Reagents

24-well tissue culture plates
Antigen
Carbonate-bicarbonate buffer, pH 9.6
Cell suspension
PBS
Tween 20 (Sigma)
PBS containing 1% BSA, 0.05% Tween 20 and 0.02% NaN_3 (PBS-Tween)
PBS
Antibodies
Dulbecco's PBS containing calcium and magnesium, with 1 % BSA
Substrate
Agarose (Type 1; Sigma)
Humidified chamber

Protocol:

1. Coat wells with 1 ml of a 2 mg/ml solution of antigen in carbonate-bicarbonate buffer. Incubate overnight at 4°C. Wash three times with PBS-Tween and twice with PBS.
2. Add 1 ml of cell solution (1 x 10^{6}cells/ml in Dulbecco's PBS with Ca^{2+}, Mg^{2+} and 1% BSA). Incubate for 2 hours at 37°C in a humidified chamber. Wash three times with PBS-Tween.
3. Add 1 ml of an appropriate dilution of antibody (use rabbit anti-mouse for mouse cells) in PBS-Tween to each well and incubate for for 2 hours at 37°C. Wash three times with PBS-Tween.
4. Add 1 ml of appropriate dilution of antibody (anti-rabbit in our case; diluted in PBS-Tween) conjugated to an enzyme. Incubate for 2 hours at 37°C. Wash three times with PBS-Tween.
5. Add 1 ml of warm (40°C) agarose-substrate mixture (3% [w/v] stock solution of warm agarose is mixed with substrate solution to a final concentration of 0.6%) to each well. Allow 1 minute for it to harden. Incubate for 30 to 60 minutes at 37°C.
6. Enumerate the macroscopic dots.

E. TWO-COLOR ELISPOT[17]

This assay is a modification of the ELISPOT technique in which a combination of two immunoenzymatic reactions, resulting in two different colors, makes it possible to measure two antigenically distinct antibodies secreted from a group of cells.

Materials and Reagents

96-well nitrocellulose-bottom Millititer HA plates (Millipore)
PBS
Paper towels
Cell suspension
Two different antibodies (e.g., anti-IgG and anti-IgA) conjugated to two
    different enzymes
Appropriate substrates
PBS
PBS containing 0.05% Tween 20 (PBS-Tween)
PBS containing 0.05% Tween 20 and 1% FCS
0.05 M Tris-buffered saline, pH 8.0
RPMI 1640 medium with 5% FCS
CO_2 incubator
Stereomicroscope

Protocol:

1. Dilute antigen in PBS to a concentration of 2 µg/ml and add 100 µl of the solution into each well of a plate. Incubate overnight at 4°C.
2. Wash three times with PBS.
3. Dry the outer surface of the nitrocellulose membrane with absorbent paper towels.
4. Fill each well with 0.2 ml of RPMI 1640 medium with 5% FCS and incubate the plate for 30 minutes at 37°C.
5. Decant the medium and add 0.1 ml of cell suspension (10^5 cells/well) in RPMI 1640 medium with 5% FCS. Incubate for 4 hours at 37°C in a CO_2 incubator.
6. Wash the plates three times with PBS and then three times with PBS-Tween. Incubate the plates in PBS-Tween for 5 minutes.
7. Decant the PBS-Tween and dry again the outer surface of the nitrocellulose membrane with absorbent paper towels.
8. Add 0.1 ml of PBS-Tween with 1% FCS and a mixture of two antibodies (e.g., anti-IgA-alkaline phosphatase and anti-IgG-horseradish

peroxidase), respectively, or vice versa, to each well.

9. Incubate for 3 hours at room temperature or overnight at 4°C.
10. Wash the plate four times with PBS and immerse in 0.05 M Tris-buffered saline, pH 8.0, for 5 minutes prior to development.
11. Decant the buffer and dry again the outer surface of the nitrocellulose membrane with absorbent paper towels.
12. Add 0.1 ml of the first substrate into each well and incubate for 15 minutes at room temperature. Wash once with PBS and dry again the outer surface of the nitrocellulose membrane with absorbent paper towels.
13. Add 0.1 ml of the second substrate into each well and incubate for an additional 10 minutes at room temperature. Rinse the plate thoroughly with tap water.
14. Allow the plate to dry and evaluate the individual wells for presence of blue and red spots under low magnification.

Comments:

1. Coated plates can be stored for several weeks at 4°C filled with coating solution, or for several months at -20°C after decanting the coating solution. Do not let them dry out.
2. Too long an incubation time in steps # 12 and # 13 might result in an excessive background staining. Check control wells without cell suspension and adjust the time adequately.

F. LABELING OF ANTIBODIES WITH ALKALINE PHOSPHATASE[18]

Materials and Reagents

Antibody
PBS
Alkaline phosphatase in NaCl solution (Sigma)
25% glutaraldehyde, EM grade (Sigma)
100 mM lysine and 100 mM ethanolamine in PBS
Blocking buffer
10-ml Sephadex G-25 column (Pharmacia)
0.2-μm filter
Small plastic tubes
p-Nitrophenyl phosphate substrate (NPP; Sigma)
NaCO$_3$
MgCl$_2$
Fraction collector

Protocol:

1. Prepare a 1:3 mixture of antibody and alkaline phosphatase in PBS at greater than 0.2 mg/ml total protein concentration.
2. Add 25% glutaraldehyde to a 0.2% final concentration while vortexing. Incubate for 2 hours at room temperature. Stop reaction by adding an equal volume of blocking buffer.
3. Desalt the antibody by passing through a Sephadex G-25 column in blocking buffer. The Sephadex G-25 volume should be 10 times larger than antibody volume. Collect fractions that are one half the antibody volume.
4. Test individual fractions by adding 2 μl into small tubes containing 0.5 ml of NPP substrate (3 mM NPP in 0.05 M $NaCO_3$ and 0.05 mM $MgCl_2$). Pool the first five fractions that strongly hydrolyze NPP.
5. Dilute the pool 1:2 in blocking buffer, filter through a 0.2-μm filter and store at 4°C for up to one year.

G. ALTERNATE METHOD FOR LABELING OF ANTIBODIES WITH ALKALINE PHOSPHATASE[19]

Materials and Reagents

Antibody
50% glutaraldehyde, EM grade (Sigma)
0.5 M Tris, pH 8
PBS
Bovine serum albumin (BSA)
NaN_3
Dialysis tube
Alkaline phosphatase in NaCl solution (Sigma)

Protocol:

1. Mix 2 mg of antibody with 5 mg of alkaline phosphatase and put the mixture into a dialysis tube. Dialyze overnight at 4°C against PBS.
2. Remove from the dialysis tube and measure the volume. Add glutaraldehyde to a final concentration of 2% (v/v). Incubate for 2 hours at 4°C with continuous mixing.
3. Transfer the mixture into a dialysis tube and dialyze overnight at 4°C against PBS.
4. Dialyze overnight at 4°C against 0.5 M Tris, pH 8.

5. Remove from the dialysis tube and dilute to 4 ml with 0.5 M Tris containing 1% BSA and 0.2% NaN$_3$.
6. Store in the dark at 4°C.

H. LABELING OF ANTIBODIES WITH HORSERADISH PEROXIDASE[20]

This method may be also used for labeling of antibodies with alkaline phosphatase.

Material and Reagents

Horseradish peroxidase (HRP; Sigma)
PBS
50% glutaraldehyde, EM grade (Sigma)
Saline
Antibody
Carbonate-bicarbonate buffer, pH 9.5
0.2 M lysine
Saturated ammonium sulfate (see Chapter 10)
Human serum albumin (HSA)
0.22-µm syringe filter
Dialysis tube

Protocol:

1. Dissolve 10 mg of HRP in 0.2 ml of PBS containing 1.25% (v/v) glutaraldehyde. Incubate overnight at room temperature with constant mixing.
2. Remove the unbreacted glutaraldehyde by dialysis overnight at 4°C against normal saline.
3. Dilute the antibody to a 5 mg/ml concentration with saline.
4. Mix 1 ml of activated HRP with 1 ml of antibody solution. Add 0.1ml of carbonate-bicarbonate buffer and incubate overnight at 4°C.
5. Add 0.1 ml of 0.2 M lysine and mix for 2 hours at room temperature.
6. Dialyze overnight at 4°C against PBS.
7. Precipitate the HRP-conjugated antibody by adding an equal volume of saturated ammonium sulfate. Spin down the precipitate by centrifugation at 2,000 x g for 20 minutes at 4°C and remove the supernatant.
8. Resuspend the precipitate in 1 ml of PBS.
9. Dialyze the conjugate for 48 hours at 4°C against two changes of PBS.

10. Centrifuge the conjugate at 10,000 x g for 30 minutes at 4°C to remove any sediment.
11. Add HSA up to 1% (w/v) and filter through a 0.22-μm syringe filter.
12. Store at -20°C.

I. LABELING OF ANTIBODIES BY A PERIODATE METHOD[21]

<u>Materials and Reagents</u>

Antibody
Alkaline phosphatase (Sigma) or horseradish peroxidase (Sigma)
0.3 M Na$_2$CO$_3$, pH 8.1
0.01 M Na$_2$CO$_3$, pH 9.5
1-Fluoro-2,4-dinitrobenzene (Sigma)
Ethanol
Bovine serum albumin (BSA)
0.08 M sodium periodate
0.16 M ethylene glycol
PBS
Sodium borohydride
Sephadex G200 (Pharmacia)
Column
Dialysis tubing
Fraction collector
Spectrophotometer

Protocol:

1. Dialyze 5 ml of alkaline phosphatase or horseradish peroxidase against 0.3 M Na$_2$CO$_3$ buffer, pH 8.1, overnight at 4°C.
2. Add 0.1 ml of a 1% solution of 1-fluoro-2,4-dinitrobenzene (dissolved in absolute ethanol) to the dialyzed solution and mix gently for 60 minutes at room temperature.
3. Add 1 ml of 0.08 M sodium periodate (freshly prepared) and gently mix for 30 minutes at room temperature.
4. Add 1 ml of 0.15 M ethylene glycol and mix gently for 60 minutes at room temperature.
5. Dialyze against 0.01 M Na$_2$CO$_3$, pH 9.5, overnight at 4°C.
6. Dialyze your antibody against 0.01 M Na$_2$CO$_3$, pH 9.5, overnight at 4°C.
7. Mix the dialyzed antibody and 3 ml of the enzyme-aldehyde solution (step # 5) and mix for 3 hours at room temperature.

197

8. Add 5 mg of sodium borohydride and incubate for 3 hours at 4°C.
9. Dialyze the conjugate against PBS for 24 hours at 4°C.
10. Apply the conjugate to the Sephadex G200 column. Elute with PBS at a flow rate of 5 ml/hour. Collect 2-ml fractions and check the absorbance at OD_{280}. The first peak contains the conjugate.
11. Store at -20°C in 1% BSA.

J. LABELING OF ANTIBODIES WITH β-D-GALACTOSIDASE[22]

Materials and Reagents

Purified antibody
β-D-galactosidase (Sigma)
N,N'-o-phenylendimaleimide (Aldrich)
Bovine serum albumn (BSA)
$MgCl_2$
1 M NaOH
0.01 M Sodium phosphate buffer, pH 7.0
0.1 M Sodium acetate buffer, pH 5.0
Sephadex G25 (Pharmacia)
Sephadex 6B (Pharmacia)
Water bath
Fraction collector
Spectrophotometer
Dialysis tubing
NaCl
NaN_3

Protocol:

1. Dissolve the antibody in 0.1 M sodium acetate buffer, pH 5.0, to a concentration of 3 mg/ml.
2. Add 2 ml of the antibody solution dropwise into 1 ml of a saturated solution of N,N'-o-phenylendimaleimide in 0.1 M sodium acetate buffer, pH 5.0, at 4°C. Incubate at 30°C for 20 minutes.
3. Equilibrate a Sephadex G25 column with 0.1 M sodium acetate buffer, pH 5.0. Apply an antibody solution and collect 1-ml fractions. Check the antibody concentration at OD_{280} and pool antibody containing fractions.
4. Adjust the pH of the isolated dimaleimide-treated antibody to 6.5 using a sodium phosphate buffer, pH 7.0.
5. Dialyze β-D-galactosidase against 0.01 M sodium phosphate buffer, pH

7.0, overnight at 4°C. Store in 0.01 M sodium phosphate buffer, pH 7.0, supplemented with 0.5% BSA and 1 mM $MgCl_2$.

5. Incubate the dimaleimide-treated antibody with β-D-galactosidase (1:1 w/w ratio) at 30°C for 20 minutes in 0.1 M sodium acetate buffer, pH 5.0.

6. Adjust the pH to 7.0 and apply to a Sepharose 6B column equilibrated with 0.01 M sodium phosphate buffer, pH 7.0, containing 0.1 M NaCl, 1 mM $MgCl_2$, 0.01% BSA, and 0.1% NaN_3.

REFERENCES

1. **Jerne, N. K. and Nordin, A. A.**, Plaque formation in agar by single antibody-producing cells, *Science,* 140, 405, 1963.
2. **Sterzl, J. and Mandel, L.**, Estimation of the inductive phase of antibody formation by plaque technique, *Fol. Microbiol.,* 9, 173, 1964.
3. **Sterzl, J. and Riha, I.**, Detection of cells producing 7S antibodies by the plaque technique, *Nature,* 208, 858, 1965.
4. **Gronowicz, E., Coutinho, A. and Melchers, F.**, A plaque assay for all cells secreting Ig of a given type or class, *Eur. J. Immunol.,* 6, 588, 1976.
5. **Henry, C. and North, J.**, Measurement of polyclonal responses, in *Selected Methods in Cellular Immunology,* Mishell, B. B. and Shiigi, S. M., Eds., W. H. Freeman and Co., San Francisco, 1980, 109.
6. **Henry, C.**, Anti-hapten plaques, in *Selected Methods in Cellular Immunology,* Mishell, B. B. and Shiigi, S., M., Eds., W. H. Freeman and Co., San Francisco, 1980, 95.
7. **Rittenberg, M. B. and Pratt, K. L.**, Antitrinitrophenyl (TNP) plaque assay. Primary response of Balb/c mice to soluble and particulate immunogen, *Proc. Soc. Exp. Biol. Med.,* 132, 575, 1969.
8. **Mishell, R. I. and Dutton, R. W.**, Immunization of normal mouse spleen cells in suspensions *in vitro, Science,* 153, 1004, 1966.
9. **Engvall, E. and Perlman, P.**, Enzyme-linked immunosorbent assay (ELISA): Quantitative assay of immunoglobulin G, *Immunochemistry,* 8, 871, 1971.
10. **Porstman, B. and Porstman, T.**, Chromogenic substrates for enzyme immunoassays, in *Nonisotopic Immunoassays,* Ngo, T. T., Ed., Plenum Press, New York, 1988, 57.

11. Feit, C., Bartal, A. H., Tauber, G., Dymbort, G. and Hirshaut, Y., An enzyme-linked immunosorbent assay (ELISA) for the detection of monoclonal antibodies recognizing antigens expressed on viable cells, *J. Immunol. Methods*, 58, 301, 1983.

12. Sedgwick, J. D. and Holt, P. G., A solid-phase immunoenzymatic technique for the enumeration of specific antibody-secreting cells, *J. Immunol. Methods*, 57, 301, 1983.

13. Czerkinsky, C., Nilsson, L. A., Nygren, H., Ouchterlony, O. and Tarkowski, A., A solid-phase enzyme-linked immunospot (ELISPOT) assay for enumeration of specific antibody-secreting cells, *J. Immunol. Methods*, 65, 109, 1983.

14. Czerkinsky, C., Anderson, G., Ferrua, B., Nordstom, I., Quiding, M., Eriksson, K., Larson, L., Hellstrand, K. and Ekre, H., Detection of human cytokine-secreting cells in distinct anatomical compartments, *Immunol. Rev.*, 119, 5, 1991.

15. Czerkinsky, C. and Sedgwick, J., Enzyme-linked immunospot (ELISPOT) assays for detection of specific antibody-secreting cells, in *Methods of Immunological Analysis*, Masseyeff, R. F., Albert, W. H. and Staines, N. A. Eds., VCH Verlagsgesellschaft, Weinheim, 1993, 504.

16. Sedwick, J., Cooke, A., Dorries, R., Hutchings, P. and Schwender, S., Detection and enumeration of single antibody- or cytokine-secreting cells by the ELISA-plaque (ELISPOT) assay, in *Laboratory Methods in Immunology, Vol. 1.*, Zola, H., Ed., CRC Press, Boca Raton, 1990, 103.

17. Czerkinsky, C., Moldoveanu, Z., Mestecky, J., Nilsson, L. A. and Ouchterlony, O., A novel two colour ELISPOT assay. I. Simultaneous detection of distinct types of antibody-secreting cells, *J. Immunol. Methods*, 115, 31, 1988.

18. Hornbeck, P. Antibody detection and preparation, in *Current Protocols in Immunology*, Coligan, J. E., Kruisbeek, A. M., Margulies, D. H., Shevach, E. M. and Strober, W., Eds., Green Publishing and Wiley-Interscience, New York, 1991, 2.0.1.

19. Avrameas, S., Coupling of enzymes to proteins with glutaraldehyde, *Immunochemistry*, 6, 43, 1969.

20. Avrameas, S. and Ternynck, T., Peroxidase labelled antibody and Fab' conjugates with enhanced intracellular penetration, *Immunochemistry*, 8, 1175, 1971.

21. Nakane, P. K. and Kawaoi, A., Peroxidase-labelled antibody; a new method of conjugation, *J. Histochem. Cytochem.*, 22, 1084, 1974.

22. Kato, K., Fukui, H., Hamaguchi, Y. and Ishikawa, E., Enzyme-linked immunoassay: conjugation of the Fab' fragment of rabbit IgG with β-D-galactosidase from *E. coli* and its use in immunoassay, *J. Immunol.*, 116, 1554, 1976.

Chapter 10

PURIFICATION OF IMMUNOGLOBULINS

Several excellent and detailed reviews and monographs describing isolation and characterization of immunoglobulins have been published.[1-3] Extinction coefficients and molecular size of immunoglobulins and their fragments are shown in Table 1.

I. PRECIPITATION BY AMMONIUM SULFATE

Materials and Reagents

Saturated $(NH_4)_2SO_4$ (SAS)
Glass beaker
Magnetic stirrer
Serum or ascites fluid
Hot plate
Dialysis tube
Glass pipette
Sodium azide
PBS
Spectrophotometer

Protocol:

1. To make SAS, dissolve about 900 g of ammonium sulfate to 1,000 ml of boiling H_2O. Let the solution cool to room temperature. Crystals of $(NH_4)_2SO_4$ will form and should always be present.
2. Clarify serum or ascites fluid by centrifugation at 10,000 x g for 30 minutes at 4°C. Decant and save the supernatant; discard membranous material and cell debris remaining in the pellet.
3. Put supernatant into a beaker and start stirring. Add SAS slowly until precipitate starts to form. Then slowly bring up to 50% (vol/vol) SAS. Allow precipatate to form for additional 2 to 4 hours at 4°C or overnight.
4. Centrifuge at 10,000 x g for 60 minutes at 4°C and save precipitate. Save the supernatant to check for residual antibody activity.

5. Dissolve precipitate in a minimum (usually less than 1/2 original volume) volume of PBS.
6. Measure the concentration of IgG at OD_{280}.
7. Dialyze against PBS (or desired buffer) overnight at 4°C. Add sodium azide to 10 mM and store at 4°C or use for further purification.

Comments:

1. Removal of lipids by a passage of ascites fluid over a funnel with glass wool might be necessary before performing step # 2.

II. CAPRYLIC ACID-AMMONIUM SULFATE ISOLATION OF IgG FROM ASCITES FLUID

Materials and Reagents

Ascites fluid
Caprylic acid (Sigma)
Fume hood
Centrifuge tubes
Saturated $(NH_4)_2SO_4$ (SAS)
PBS
Dialysis tube
Sodium azide
Acetic acid
Stirring plate
pH meter

Protocol:

1. Adjust the pH of the ascites fluid to 4.95 using 3 M acetic acid. The right pH is extremely important.
2. With stirring at room temperature under the fume hood, slowly add caprylic acid to a final concentration of 5% (1:20).
3. Stir for 30 min at room temperature.
4. Centrifuge at 10,000 x g for 30 min at 20°C.
5. Supernatant contains the antibody; thus save supernatant and discard precipitate.
6. Measure volume of the supernatant.

7. Slowly add an equal volume of SAS while stirring at room temperature.
8. Stir for two hours at 4°C.
9. Centrifuge as in step # 4.
10. Decant supernatant and discard; save precipitate.
11. Blot tube well to remove as much ammonium sulfate as possible.
12. Redissolve precipitate in a minimum volume of PBS (usually less than 1/2 volume of original ascites).

TABLE 1

Extinction Coefficients and Molecular Size of Immunoglobulins and Their Fragments

Molecule	A_{280} (1 mg/ml)	Molecular Weight (kDa)		
		Nonreduced	Reduced	
			Heavy chain	Light chain
IgG	1.43	160	55	25
IgM	1.18	900	78	25
Fab	1.53	50	25	25
$F(ab)_2$	1.48	100-110	25	25
$F(ab)_2\mu$	1.38	135	44	25
$F(ab)\mu$	1.38	65	44	25

13. Dialyze overnight against PBS (or other desired buffer) at 4°C to remove residual ammonium sulfate.
14. Centrifuge again as in step # 4 to remove undissolved precipitate.
15. Either add sodium azide to 10 mM and store the antibody at 4°C or freeze at -80°C.

Comments:

1. Use polycarbonate, polysulfone, nylon, or glass centrifuge tubes because the caprylic acid will dissolve polyethylene, polystyrene, and polypropylene plastic tubes.

III. PURIFICATION OF ANTIBODIES BY PRECIPITATION WITH POLYETHYLENE GLYCOL[4,5]

Different monoclonal antibodies precipitate out at different concentrations of polyethylene glycol (PEG). Generally, the higher the concentration of the PEG that is used, the higher the resulting contamination of the sample. Some IgG and most IgM antibodies precipitate out below 5% PEG.

Materials and Reagents

Polyethylene glycol 6,000 (PEG)
Ascites fluid
Veronal buffered saline (VBS)
Centrifugation tubes
Eppendorf microtubes
Electrophoresis equipment

Protocol:

1. Clarify ascites fluid by centrifugation at 10,000 x g for 30 minutes at 4°C. Decant and save the supernatant; discard membranous material and cell debris remaining in the pellet.
2. Prepare stock solutions (w/v) of 0%, 4%, 8%, 12%, 16%, 20%, 24%, 28%, 32%, and 50% PEG in VBS. Store them at 4°C.
3. In order to determine the optimal concentration of PEG for a particular antibody, perform a pilot test. Put 20 µl of cold clarified ascites fluid into Eppendorf microtubes and add 20 µl of different concentrations of PEG.
4. Mix the tubes and incubate them for 30 minutes on ice. Spin the tubes at 2,000 x g for 20 minutes at 4°C.
5. Test the supernatant for the presence of the monoclonal antibody band. The concentration of PEG that removed most of the antibody band is the optimal concentration.

6. Perform the same experiment on a large scale using only the optimal concentration of PEG.

Comments:

1. Step # 6 might be repeated to achieve higher purity. Unlike ammonium sulfate precipitation, multiple precipitations in PEG and redissolving do not affect their functional characteristics.

IV. PURIFICATION OF ANTIBODIES FROM CULTURE SUPERNATANTS BY PRECIPITATION WITH POLYETHYLENE GLYCOL[4]

<u>Materials and Reagents</u>

Polyethylene glycol 6,000 (PEG)
Tissue culture supernatant
Veronal buffered saline (VBS)
Centrifugation tubes
Electrophoresis
0.3 M phosphate buffer, pH 6.8

Protocol:

1. Clarify culture supernatant by centrifugation at 10,000 x g for 30 minutes at 4°C. Decant and save the supernatant; discard membranous material and cell debris remaining in the pellet.
2. Add solid PEG at 12.5% (w/v) concentration to cold clarified culture supernatant.
3. Stir until all the PEG is dissolved.
4. Incubate overnight at 4°C.
5. Collect the precipitate by centrifugation at 2,000 x g for 20 minutes at 4°C.
6. Dissolve the precipitate in 2% of the original volume VBS. Incubate for 2 hours at 4°C.
7. Centrifuge at 2,000 x g for 20 minutes at 4°C.
8. Collect the supernatant. Dissolve the pellet with 1% original volume of culture supernatant with 0.3 M phosphate buffer, pH 6.8.
9. Centrifuge at 2,000 x g for 20 minutes at 4°C.
10. Test for presence of antibodies by electrophoresis.
11. Isolate the monoclonal antibodies by performing steps # 3 to 6 from previous method.

Comments:

1. Serum-free tissue culture supernatants cannot be purified using this method due to the low concentration of proteins, which does not allow effective PEG precipitation.

V. AFFINITY CHROMATOGRAPHY USING PROTEIN A-SEPHAROSE[1,6]

Materials and Reagents

Tissue culture supernatant or ascites fluid
1 M NaOH
Protein A-Sepharose CL-4B (Pharmacia)
0.1 M citric acid
3 M potassium thiocyanate
1.5 x 10-cm glass column
Fraction collector
Dialysis tubing
PBS, pH 8.0
Spectrophotometer

Protocol:

1. Clarify culture supernatant or ascites fluid by centrifugation at 10,000 x g for 30 minutes at 4°C. Decant and save the supernatant; discard membranous material and cell debris remaining in the pellet.
2. Dilute ascites fluid 10x with PBS, pH 8.0 or adjust tissue culture supernatant to pH 8.0 by dialysis against PBS, pH 8.0, overnight at 4°C.
3. Prepare Protein A-Sepharose column and attach a fraction collector.
4. Equilibrate column with PBS, pH 8.0, at 4°C.
5. Load the column with supernatant/ascites fluid.
6. Wash the column with several volumes of PBS, pH 8.0.
7. Check OD_{280} of the eluate. Keep washing until the OD_{280} is at the baseline.
8. Elute IgG with 0.1 M citric acid at pH optimal for desired isotype. For mouse IgG use pH 6.5, for IgG_{2a} use pH 4.5, and for IgG_{2b} and IgG_3 use pH 3.0.
9. Pool fractions containing immunoglobulins (check OD_{280}). Put the eluates into dialysis tubing and dialyze against PBS, pH 7.3, with 0.02% NaN_3 overnight at 4°C.

10. Clean the column with 1 volume of filtered 3 M potassium thiocyanate. Reequilibrate the column in PBS by several volumes of PBS, pH 7.3. Store at 4°C.

Comments:

1. Most mouse and human IgG antibodies bind to Protein A.
2. The binding capacity of a Protein A-Sepharose column is approximately 5 mg mouse IgG, and 8 mg human IgG, per 1 ml of Protein A-Sepharose.

VI. PURIFICATION OF IgM[1,2]

Materials and Reagents

IgM
Borate buffer
Dialysis tube
Size-exclusion (SE) column
SDS-PAGE electrophoresis
Fraction collector

Protocol:

1. Dialyze antibody against H_2O at 4°C for 24 hours.
2. Centrifuge dialyzed antibody at 20,000 x g at 4°C for 60 minutes. Discard supernatant and dissolve pellet in 5 ml borate buffer.
3. Prepare an SE column and equilibrate with borate buffer. Add antibody and elute with borate buffer. Collect 100 fractions. IgM will be in the first protein peak.
4 Measure IgM concentration at OD_{280}.
5. Check the purity of the IgM on a 10% SDS-PAGE gel under reducing conditions.
6. Store in borate buffer at 4°C or at -70°C.

Comments:

1. Most IgM antibodies are not soluble in water.

VII. AMMONIUM SULFATE PRECIPITATION OF IgM[1,2]

Materials and Reagents

Saturated $(NH_4)_2SO_4$ (SAS)
Glass beaker
Magnetic stirrer
Serum or ascites fluid
Spectrophotometer
Size-exlusion (SE) column
SDS-PAGE electrophoresis
Borate buffer
Fraction collector

Protocol:

1. Clarify serum or ascites fluid by centrifugation at 10,000 x g for 30 minutes at 4°C. Decant and save the supernatant; discard membranous material and cell debris remaining in the pellet.
2. Put supernatant into a beaker and start stirring. Add SAS slowly to 45% (v/v) of antibody). Allow precipitate to form for additional 2 to 4 hours or overnight at 4°C .
3. Centrifuge at 10,000 x g for 60 minutes at 4°C and save precipitate. Save the supernatant to check for residual antibody activity.
4. Dissolve precipitate in a minimum (usually less than 1/2 original volume) volume of borate buffer.
5. Prepare an SE column and equlibrate with borate buffer. Add antibody and elute with borate buffer. Collect 100 fractions. IgM will be in the first protein peak.
6 Measure IgM concentration at OD_{280}.
7. Check the purity of the IgM on a 10 % SDS-PAGE gel under reducing conditions.
8. Store in borate buffer at 4°C or at -70°C.

Comments:

1. Use for IgM antibodies that do not precipitate in water.

VIII. PURIFICATION OF ANTI-HAPTEN ANTIBODIES BY AFFINITY CHROMATOGRAPHY

This procedure describes purification of anti-Arsonate (Ars) antibodies.[7] You can either activate agarose with CNBr by yourself[8] or use commercially available CNBr-activated agarose.

Materials and Reagents

Stirring plate
CNBr-activated agarose (Pharmacia)
$NaHCO_3$ buffer
Glass beaker
Parafilm™
Sintered glass funnel
PBS
H_2O
Acetic acid
Column
Sodium azide
Antiserum
Spectrophotometer

Protocol:

1. Add 1 ml of Ars (40 mM solution in 0.1 M $NaHCO_3$, pH 9.0 buffer) to a 10 ml CNBr-activated agarose in glass beaker. Cover the beaker with Parafilm and stir the mixture slowly for 24 hours at 4°C.
2. Transfer the CNBr-activated agarose to a sintered glass funnel and wash with 1 liter of ice-cold H_2O followed by 1 liter of ice cold 0.1 M $NaHCO_3$, pH 9.0 buffer. Resuspend CNBr-activated agarose in PBS.
3. Put CNBr-activated agarose into a suitable column. Slowly add the antiserum to the column and collect the eluate. Ten ml of CNBr-activated agarose will bind approximately 200 mg of specific antibody.
4. Wash the column with PBS until the eluate has an OD_{280} of less than 0.01.
5. Combine all eluted fractions, concentrate and dialyze against PBS overnight at 4°C. Determine the amount of antibody by absorbance at 280 nm.

211

Comments:

1. The column can be reused. It is necessary to remove all antibodies from the column by washing with 2 volumes of 0.1 M acetic acid before reuse. Equilibrate the column with 2 liters of PBS and store at 4°C with 0.02% sodium azide.

IX. FRAGMENTATION OF IMMUNOGLOBULIN INTO F(ab) FRAGMENTS USING PAPAIN

Papain digests specifically IgG into F(ab) fragments. This technique can be used for every subclass of IgG molecule.

Materials and Reagents

2 mg/ml purified antibody in PBS
Papain (2X crystallized suspension; Sigma # P3125)
Digestion buffer
PBS
0.3M iodoacetamide (prepare fresh from crystalline material) in PBS
Eppendorf microcentrifuge tubes
Water bath
SDS-PAGE electrophoresis
Dialysis tube
Glass Pasteur pipette
Burner
Spectrophotometer

Protocol:

1. Determine protein concentration of purified antibody (O.D. at 280 nm).
2. Pipette 100 µl of 2 mg/ml purified IgG in PBS into each of 13 numbered microcentrifuge tubes.
3. Prepare two papain solutions in 5 ml digestion buffer: one at 0.02 mg/ml papain and the other at 0.04 mg/ml.
4. Add 100 µl of 0.02 mg/ml papain to 6 of the IgG tubes. Add 100 µl of 0.04 mg/ml papain to a second group of 6 IgG tubes. Finally, add 100 µl of digestion buffer without papain to 1 of the IgG tubes.
5. Prepare a digestion time curve by incubating tubes in a 37°C water bath and removing them at different times in groups of two (one tube each of enzyme:antibody ratio at 1:20 and 1:100). Remove the first group after 1 hour and subsequent groups at 2, 4, 5, 6, and 24 hours. Remove

the control tube with the 24 hour set. As each group is removed from the water bath, terminate the reaction by adding 20 μl of 0.3 M iodoacetamide in PBS and then vortexing.

6. Puncture the lid of another set of microcentrifuge tubes using a heated glass Pasteur pipette (wide end heated) to completely remove the center part of the lid. Pipette contents of reaction tubes into similarly labeled dialysis tubes. Open the lid and place a single layer of dialysis tubing over the top of the tube, then close lid to hold dialysis tubing in place.

7. Turn tube upside down and shake reaction mixture onto the membrane surface. Tape each tube, dialysis surface down, to the side of a beaker filled with PBS and dialyze 4 hours at 4°C with stirring.

8. Analyze digestion products by running 12.5% nonreducing SDS-PAGE electrophoresis.

X. FRAGMENTATION OF IMMUNOGLOBULIN INTO F(ab) FRAGMENTS USING PREACTIVATED PAPAIN[1]

This technique is a gentler method; thus, it is not necessary to perform a pilot experiment.

Materials and Reagents

10 mg IgG in 5 ml PBS
Acetate/EDTA buffer
2 mg/ml papain acetate/EDTA buffer
0.05 M cysteine (free base, Sigma)
Iodoacetamine crystals
PD-10 column (Pharmacia)
Dialysis tube
Water bath
PBS, pH 8.0
Protein A-Sepharose column
Size exclusion (SE) column
SDS-PAGE electrophoresis
PBS containing 0.02% NaN$_3$
Spectrophotometer
Fraction collector
Amicon concentrator

Protocol:

1. Dialyze IgG in acetate/EDTA buffer at 4°C overnight.

2. Determine the concentration of antibody at OD_{280}.
3. Incubate 2 mg/ml papain and 0.05 M cysteine for 30 minutes in a 37°C water bath.
4. Equilibrate a PD-10 column with 20 ml of acetate/EDTA buffer.
5. Apply papain/cystein mixture to the PD-10 column. Elute with acetate/EDTA buffer and collect ten 1-ml fractions.
6. Check the fractions at OD_{280} and pool 3 fractions containing papain. Calculate the preactivated papain concentration:

$$A_{280}/2.5 = \text{mg preactivated papain.}$$

7. Add 0.5 mg preactivated papain to purified IgG in acetate/EDTA buffer. Shake and incubate 10 hours in a 37°C water bath.
8. Add crystalline iodoacetamine to a final concentration of 0.03 M.
9. Dialyze against PBS, pH 8.0, at 4°C overnight.
10. Calibrate a Protein A-Sepharose column with PBS, pH 8.0. Add papain/IgG mixture. Collect unbound 2-ml fractions. Pool fractions forming the first peak and concentrate to a volume lower than 5 ml.
11. Load the sample on an SE column. Collect 100 fractions and check the molecular weight of fractions using SDS-PAGE electrophoresis.
12. Store in PBS containing 0.02% NaN_3 at -70°C.

XI. FRAGMENTATION OF IMMUNOGLOBULIN INTO F(ab')$_2$ FRAGMENTS USING PEPSIN

Pepsin is used for a fragmentation of IgG into F(ab')$_2$ fragments. The optimal pH for pepsin is around 2, but since it would be damaging for the antibodies, a pH of 4.1 is used. Do not use this method for F(ab')$_2$ fragmentation of IgG$_{2b}$, as Fab/c fragments will be produced.

Materials and Reagents

Purified antibody
0.1 M sodium acetate buffer, pH 4.1
Dialyzing tubes
Pepsin (Sigma)
1 M Trizma base
0.5 M NaCl
0.05 M phosphate buffer, pH 7.3
Amicon concentrator
Eppendorf microcentrifuge tubes

Water bath
Electrophoresis
S-200 (Pharmacia)
Fraction collector
Spectrophotometer

Protocol:

1. Dialyze purified antibody overnight against 0.1 M sodium acetate buffer, pH 4.1, at 4°C.
2. Determine protein concentration of purified antibody (O.D. at 280 nm).
3. Dilute antibody to 5 mg/ml in 0.1 M sodium acetate buffer. Put 0.2 ml of the diluted antibody into Eppendorf microcentrifuge tubes.
4. Prepare fresh pepsin at an exact concentration of 1 mg/ml in 0.1 M sodium acetate buffer, pH 4.1.
5. Use pepsin-to-antibody percentages of:
   A. 0.1% (0.1 µl of pepsin to 100 µg/ml of antibody)
   B. 0.2% (0.2 µl of pepsin to 100 µg/ml of antibody)
   C. 1.0% (1.0 µl of pepsin to 100 µg/ml of antibody)
   D. 2.0% (2.0 µl of pepsin to 100 µg/ml of antibody)
   E. 5.0% (5.0 µl of pepsin to 100 µg/ml of antibody)
6. Digest for 7 hours at 37°C.
7. Add an equal volume of 1 M Trizma base into each tube to stop digestion.
8. Run an SDS electrophoresis to determine the percentage of F(ab')$_2$ fragments.
9. Use the optimal dose of pepsin on a large-scale experiment.
10. Reconcentrate to 10 mg/ml on an Amicon concentrator.
11. Equilibrate an S-200 column of around 1 m length with 0.05 M phosphate buffer, pH 7.3, and 0.5 M NaCl.
12. Run your sample (up to 15 ml) through the column.
13. Collect, monitor (O.D. at 280 nm), and plot fractions. The F(ab')$_2$ will elute following the undigested antibody.

Comments:

1. The pH of 0.1 M sodium acetate buffer is very critical.
2. The optimal concentration of pepsin depends on several factors such as the antibody, activity of the pepsin, etc. Therefore, it is necessary to perform a pilot experiment prior to a large-scale digestion.

XII. DOUBLE IMMUNODIFFUSION ASSAY[9]

The double immunodiffusion assay is a simple method for detecting the presence of antigen-specific antibodies. Solutions of antigen and antibody are placed in adjacent wells and allowed to diffuse into the gel. In case of an antigen-antibody reaction, a immunoprecipitation line will form between the wells containing antigen and antigen-specific antibody.

Materials and Reagents

PBS
Antisera
Antigen
Staining solution
Destaining solution
Agarose
Microscope slides
Ouchterlony template
Aspirator
Coplin jar
Filter paper
37°C incubator

Protocol:

1. Melt 1% agarose and apply no more than 3.3 ml of agarose to the slide (keep slides on a level surface).
2. Allow agarose to harden, but do not allow to dry. If the slide is to be out awhile, place it in a wet box.
3. Place slide on an Ouchterlony template and punch holes in the agarose.
4. Carefully remove the slide from the template. Remove the agarose plugs with the aspirator.
5. Make serial dilutions of the antigen in PBS.
6. Place approximately 12 μl of antigen in each of the outer wells. The volume will vary according to how much agarose was applied to the slide. Fill each well to the top.
7. In the center well, add approximately 25 μl of antibody (also filling it to the top).
8. Place the slides in the wet box and incubate it at 4°C for 2 days. Initial results may be seen in just a few hours, but to develop completely the slides should be allowed to develop for 2 days.

9. After the 48-hour incubation, place slides in a Coplin jar filled with PBS for 2 days.
10. Remove slides from PBS and place filter paper over the agarose, wrapping the edges of the filter paper around the slide. Place in 37°C incubator and bake for approximately 2 hours to dry the agarose.
11. Stain slides in staining solution for 10 minutes.
12. Destain for 4 minutes in destaining solution. Remove slides and place in fresh destaining solution for another 4 minutes.

REFERENCES

1. **Andrew, S. M. and Titus, J. A.,** Purification of antibodies and preparation of antibody fragments, in *Current Protocols in Immunology,* Colignan J. E., Kruisbeek A. M., Margulies D. H., Shevach E. M., and Strober W., Eds., Green Publishing and Wiley-Interscience, New York, 1992, 2.7.1.
2. **Hardy, R. R.,** Purification and characterization of monoclonal antibodies, in *Handbook of Experimental Immunology, Vol.1, Immunochemistry,* Weir D. M., Eds., Blackwell Scientific, Oxford, 1986, 13.1.
3. **Kabat, E. A.,** *Structural Concepts in Immunology and Immunochemistry,* 2nd ed. Holt, Reinhart & Wilson, New York, 1976.
4. **Neoh, S. and Zola, H.,** Purification of mouse monoclonal antibodies from ascitic fluid and culture supernatants by precipitation with polyethylene glycol 6000, in *Laboratory Methods in Immunology,* Zola H., Ed., CRC Press, Boca Raton, 1990, 73.
5. **Neoh, S. H., Gordon, C., Potter, A. and Zola, G.,** The purification of mouse monoclonal antibodies from ascitic fluid, *J. Immunol. Meth.,* 91, 231, 1986.
6. **Ey, P. L., Prouse, S. J. and Jenkin, C. R.,** Isolation of pure IgG_1, IgG_{2a} and IgG_{2b} immunoglobulins from mouse serum using protein A-Sepharose, *Immunochemistry,* 15, 429, 1978.
7. **Wofsy, L. and Burr, B.,** The use of affinity chromatography for the specific purification of antibodies and antigens, *J. Immunol.,* 103, 380, 1969.

8. Hood, A. H., Wofsy, L., Kimura, J. and Henry, C., Purification of immunoglobulins and their fragments, in *Selected Methods in Cellular Immunology*, Mishell B. B. and Shiigi S. M., Eds., W. H. Freeman, San Francisco, 1980, 278.

9. Ouchterlony, O. and Nilsson, L. A., Immunodiffusion and immunoelectrophoresis, in *Handbook of Experimental Immunology*, Vol. *1.*, *Immunochemistry*, Weir D. M., Herzenberg L. A., and Blackwell C., Eds., Blackwell, Oxford, 1986, 32.1.

Chapter 11

BUFFERS AND MEDIA

ACK lysing buffer

NH_4Cl	8.29 g
$KHCO_3$	1 g
Na_2EDTA	37.2 mg

Add 800 ml H_2O and adjust pH to 7.2 with 1 N HCl. Add H_2O to 1000 ml. Filter through a 0.2-μm filter and store at room temperature.

Hemolytic Gey's solution

Stock A:

NH_4Cl	35 g
KCl	1.85 g
$Na_2HPO_4 \cdot 12\ H_2O$	1.5 g
KH_2PO_4	0.12 g
Glucose	5 g
Phenol red	60 mg
H_2O	1,000 ml

Autoclave.

Stock B:

$MgCl_2$	0.42 g
$MgSO_4$	0.14 g
$CaCl_2$	0.34 g
H_2O	1,000 ml

Autoclave.

Stock C:

NaHCO₃	2.25 g
H₂O	100 ml

$$NaHCO_3 \quad 2.25\ g$$
$$H_2O \quad 100\ ml$$

Autoclave.

1 x Gey's solution

20 parts Stock A
5 parts Stock B
5 parts Stock C
70 parts sterile H₂O

Tris-buffered ammonium chloride

Stock A:

NH₄Cl	8.3 g
H₂O	1,000 ml

Stock B:

Tris base: dissolve 20.6 g Tris base in 900 ml water, adjust pH to 7.65 with HCl. Add water to 1,000 ml.

Working solution:

Mix 90 ml of Stock A with 10 ml of Stock B, adjust pH to 7.2

Glucose-phosphate-buffered saline (G-PBS)

Dextrose	10 g
NaCl	8.2 g
1.0 M phosphate buffer, pH 7.6	10 ml
H₂O	1,000 ml

1.0 M Phosphate buffer, pH 7.6

Stock A:

KH₂PO₄	136.1 g

H_2O	1,000 ml

Stock B:

$Na_2HPO_4 \cdot 7 H_2O$	268.1 g/l
H_2O	1,000 ml

(Warm to 45°C to dissolve.)

Mix 90 ml of Stock A with 910 ml of Stock B. Adjust to pH 7.6 with Stock B.

0.3 M Phosphate buffer

KH_2PO_4	40.7 g	270 ml
$Na_2HPO_4 \cdot 7 H_2O$	80.4 g	930 ml

Adjust to pH 6.8 and store at 4°C.

Borate-buffered saline (BBS)

Put the following ingredients into a 4-l glass beaker and heat on stirrer hot plate:

NaCl	43.9 g
H_3BO_3	61.9 g
H_2O	5,000 ml

Adjust to pH 8.0 with 0.15 M NaOH. Add H_2O up to a total volume of 6 liters.

Borate-buffered saline (BBS) (for Biotin label)

Put the following ingredients into a 4-liter glass beaker and heat on stirrer hot plate (for 6 liters):

Na tetraborate·10 H_2O	57.24 g
NaCl	26.28 g
H_3BO_3	37.1 g

After everything is dissolved, add to 8-liter carboy and fill to 6 liters with H_2O. Adjust pH to 8.43 and filter before using. Store at room temperature.

Percoll mix solution

10 x PBS	45 ml
0.6 N HCl	3 ml
H_2O	132 ml

Adjust to pH 7.0 to 7.2.

Türk solution

Acetic acid, concentrated	3 ml
1 % crystal violet in H_2O	3 ml
H_2O	294 ml

DMEM medium

DMEM with high glucose

MOPS [3-(N-morpholino)propanesulfonic acid]	2.09 g
0.05 mM 2-mercaptoethanol	
L-glutamine	216 mg
L-arginine	116 mg
L-asparagine	36 mg
Folic acid	6 mg

Add H_2O up to a total volume of 1,000 ml.

Labeling buffer

Put the following ingredients into a 4-liter glass beaker and heat on stirrer hot plate. After dissolution, add to 20-liter carboy and fill to 10 liters with H_2O.

Boric acid	30.9 g
NaCl	116.88 g

Adjust pH to 9.2 and store at 4°C.

Dialyzing buffer

TRIS-HCl	157.6 g
NaN_3	10 g

NaCl	116.88 g

Adjust pH to 7.4 and store at 4°C.

Carbonate-bicarbonate buffer

Na_2CO_3	1.59 g
$NaHCO_3$	2.93 g
NaN_3	0.2 g
H_2O	1,000 ml

Adjust pH to 9.6. Store at 4°C for no longer than 2 weeks.

5x EGTA buffer

EGTA	7.61 g
KCl	8.95 g
H_2O	1,000 ml

Store at 4°C.

Calcium-EGTA buffers

5x EGTA buffer
$CaCO_3$
1 M $CaCl_2$, pH 8.0
1 M HEPES
1 M HCl
120 mM KCl

1. Add $CaCO_3$ to 20 ml of 5x EGTA buffer. Add enough $CaCO_3$ to titrate 80% of the EGTA. Heat and stir until dissolved. Adjust to pH 9.2 with 1 M HCl. This buffer is 4x calcium/5x EGTA.
2. Slowly add 1 M $CaCl_2$ to 4 ml of 4x calcium/5x EGTA with constant measuring of the pH levels at room temperature. Stop adding $CaCl_2$ when pH stops decreasing. Calculate the number of millimoles of $CaCl_2$ were required and add the appropriate amount of 1 M $CaCl_2$ to 10 ml of 4x calcium/5x EGTA. This buffer is 5x calcium/EGTA buffer.
3. Add 1 M HEPES to 20 mM final and 1 M $MgCl_2$ to 1 mM final to the 5x calcium/EGTA buffer. Adjust to the desired pH with 1 M HCl and bring to 50 ml with 120 mM KCl. This buffer is 1x

calcium/EGTA (K_2CaEGTA) buffer.
4.   Add 1 M HEPES to 20 mM final and 1 mM $MgCl_2$ to 1 mM final to 10 ml of 5x EGTA buffer. Adjust to the same pH as previous buffer and add 120 mM KCl up to 50 ml. This buffer is 1x EGTA (K_2H_2EGTA) buffer.
5.   To obtain a buffer of a desired Ca^{2+} concentration, determine the required amounts of total calcium and EGTA and mix the 1x EGTA and calcium/EGTA buffers in the appropriate ratio.

Hanks' balanced salt solution with calcium and magnesium

Hanks' balanced salt solution

$CaCl_2$	0.11 g
$MgCl_2$	95.21 mg
FCS	10 ml
H_2O	1,000 ml

Store at 4°C.

Veronal buffered saline

NaCl	8.77 g
$MgCl_2$	76.17 mg
$CaCl_2$	33.3 mg
Sodium barbital	8.24 mg
NaN_3	0.2 g
H_2O	1,000 ml

Adjust pH to 7.4 and store at 4°C.

Blocking buffer

Borate buffered saline containing:

EDTA	0.292 g
Tween 20	0.5 ml
BSA	2.5 g
NaN_3	0.5 g
$MgCl_2$	0.238 g
H_2O	1,000 ml

Cacodylate buffer

Cacodylic acid   38.6 g

Add 900 ml H_2O and adjust pH to 6.9 with 10 M NaOH. Bring volume to 1,000 ml and store at 4°C. Cacodylic acid is highly poisonous.

Staining solution

0.5% Coomassie Brilliant Blue R-250
40% ethanol
10% glacial acetic acid
50% H_2O

Store at room temperature.

Destaining solution

15% ethanol
5% glacial acetic acid
80% H_2O

Store at room temperature.

0.1 M Sodium acetate buffer

Stock A:

Sodium acetate	51.7 g
H_2O	750 ml

Stock B:

Glacial acetic acid	21.7 ml
H_2O	728.3 ml

Stock C:

3 M NaCl	1,000 ml

225

Lower the pH of the solution A with solution B to pH 4.1. Mix 200 ml of solution A with 30 ml of solution C and 770 ml of H_2O. Readjust the pH with solution B to pH 4.1

Acetate/EDTA buffer

3 mM EDTA-HCl	0.9 g
0.1 M sodium acetate	13.6 g
H_2O	1,000 ml

Adjust to pH 5.5 with glacial acetic acid.

Borate buffers, pH 8.5, 9.0 and 10.0

0.1 M, pH 8.5

Boric acid	18.55 g
H_2O	3,000 ml

0.16 M, pH 9.0

Sodium borate ($Na_2B_4O_7$)	6.1 g
NaCl	0.76 g
H_2O	100 ml

0.1 M, pH 10.0

Boric acid	0.618 g
H_2O	100 ml

0.1 M Diethanolamine buffer, pH 9.8

0.2 M Diethanolamine	50 ml
0.2 M HCl	5.74 ml
H_2O	44.26 ml

DTNB solution

Dissolve 39.6 mg DTNB in 10 ml of 0.05 M phosphate buffer, pH 8.0. Adjust to pH 8.0 with 0.1 M NaOH if necessary. Prepare fresh for each assay.

0.3% Glutaraldehyde

25% glutaraldehyde	1.2 ml
Borate buffer, pH 10.0	98.8 ml

Prepare immediately before use.

0.01 M Phosphate buffer, pH 7.0

KH_2PO_4	1.36 g
H_2O	to 1,000 ml

Adjust pH with 10 M NaOH.

0.05 M Phosphate buffer, pH 6.0 and 8.0

KH_2PO_4	6.8 g
H_2O	to 1,000 ml

Adjust pH with 10 M NaOH.

0.05 M Tris-buffered saline

0.05 M Tris-HCl	0.8 g
1.5 M NaCl	9 g
H_2O	100 ml

Adjust to pH 8.0 and store at room temperature.

Dulbecco's PBS with Ca^{2+}, Mg^{2+} and BSA

Dulbecco's PBS	990 ml
$CaCl_2$	0.11 g
$MgCl_2$	95.21 mg
BSA	10 ml

Balanced salt solution (BSS)

The 10X BSS is made up as two stock solutions.

Stock # 1:

Dextrose	10 g
KH_2PO_4	0.6 g
$Na_2HPO_4 \cdot 7 H_2O$	3.58 g
0.5% phenol red solution	20 ml

Dissolve and bring up to 1,000 ml with distilled water.

Stock # 2:

$CaCl_2 \cdot 2 H_2O$	1.86 g
KCl	4 g
NaCl	80 g
$MgCl_2$, anhydrous	1.04 g
$MgSO_4 \cdot 7 H_2O$	2 g

Dissolve and bring up to 1,000 ml with distilled water.

1. Test the 10X stocks by making a sample of 1X BSS. Mix 10 ml of stock # 1 and 10 ml of stock # 2 and bring up to 100 ml with distilled water. The 1X BSS should be at pH 7.2 to 7.4 and have a conductivity of 14 to 16 mS.
2. 2X and 1X BSS are obtained by appropriate dilutions of the 10X stocks with distilled water. The 2X and 1X BSS can be sterilized by passage through a 0.22-µm filter.

Comments:

1. Store the 10X stock at 4°C.
2. Make nonsterile 1X BSS for the hemolytic assays just prior to use. Discard the rest of 1X BSS after use.
3. Prepare sterile 2X and 1X BSS immediately after the 10X stocks are made. Then incubate the filtered BSS at 37°C for one week prior to use to detect bacterial contamination. The sterile BSS is stored at room temperature and appears to have an indefinite shelf life.

Chapter 12

COMMERCIAL SOURCES

Accurate Chemical and Scientific
300 Shames Drive
Westbury, NY 11590
Tel: (800) 645-6264
Fax: (516) 997-4948

Advanced Magnetics, Inc.
61 Mooney St.
Cambridge, MA 02138-1038
Tel: (800) 343-1346
Fax: (617) 497-6927

Aldrich Chemical Co., Inc.
940 W. St. Paul Avenue
Milwaukee, WI 53233
Tel: (800) 558-9160
(414) 273-3850
Fax: (800) 962-9591
(414) 273-4979

Amac, Inc.
160B Larrabee Rd.
Westbrook, ME 04092
Tel: (800) 458-5060
Fax: (207) 854-0116

Amersham Corporation
2636 Clearbrook Drive
Arlington Heights, IL 60005
Tel: (800) 323-9750
(708) 593-6300
Fax: (800) 228-8735
(708) 437-1640

Amgen Biologicals
Amgen Center
Thousand Oaks, CA 91320
Tel: (800) 343-7475
Fax: (805) 498-9377

Amicon Corporation
Scientific Systems Division
21 Hartwell Avenue
Lexington, MA 02173
Tel: (617) 862-7050

ATCC (American Type Culture Collection)
12301 Parklawn Drive
Rockville, MD 20852-1776
Tel: (800) 638-6597
(301) 881-2600
Fax: (301) 231-5826

Becton Dickinson Immunocytometry Systems
2350 Qume Drive
San Jose, CA 95131
Tel: (800) 235-5953
Fax: (408) 954-2009

Bio-Rad Laboratories
2200 Wright Avenue
Richmond, CA 94804
Tel: (415) 234-4130

Biosource International
950 Flynn road, Unit A
Camarillo, CA 93012
Tel: (800) 242-0607
Fax: (805) 987-3385

Biotech, C-C Biotech Corporation
16766 Espola Road
Poway, CA 92064
Tel: (619) 451-9949
Fax: (619) 487-8138

Biotex Laboratories Inc.
#100, 8905-51 Avenue
Edmonton, Alberta, Canada T6E 5J3
Tel: (800) 661-1426
(403) 448-0621
Fax: (403) 448-0624

BioWhittaker, Inc.
8830 Biggs Ford Rd.
Walkersville, MD 21793
Tel: (800) 638-8174
(301) 898-7025
Fax: (301) 845-8338

Boehringer-Mannheim
9115 Hague Road
P.O. Box 50414
Indianapolis, IN 46250-0414
Tel: (800) 262-1640
Fax: (800) 845-7355, ext. 9409, use handset
(317) 576-2754

Calbiochem-Novabiochem Corporation
P.O. Box 12087
La Jolla, CA 92039-2087
Tel: (800) 854-3417
(619) 450-9600
Fax: (800) 776-0999
(619) 453-3552

Cedarlane Accurate Chemicals & Scientific Corp.
300 Shames Dr.
Westburg, NY 11590
Tel:(800) 645-6264
Fax: (516) 997-4948

Cistron Biotechnology Inc.
10 Bloomfield Avenue
Pine Brook, NJ 07058
Tel: (800) 642-0167
Fax: (201) 575-4854

Cleveland Scientific
P.O. Box 300
Bath, OH 44210
Tel: (216) 666-7676
Fax: (216) 666-2240

Corning Glass Works
Customer Service
Corning Science Products
P.O. Box 5000
Corning, NY 14830
Tel: (800) 222-7740
(607) 974-4667
Fax:(607) 974-7919

Costar
Corporate Headquarters
One Alewife Center
Cambridge, MA 02140
Tel: (800) 492-1110
(617) 868-6200
Fax: (617) 868-2076

Coulter Cytometry
440 W. 20th Street
Hialeah, FL 33010
Tel: (800) 633-7427
Fax: (305) 883-6899

Curtin Matheson Scientific Co. (CMS)
General offices
P.O. Box 1546
Houston, TX 77251-154
Tel: (713) 820-9898
Fax: (713) 878-2444

Difco Laboratories, Inc.
P.O. Box 1058A
Detroit, Michigan 48640
Tel: (313) 961-0800
Fax: (313) 462-8517

Du Pont NEN Research Products
Customer Service
549 Albany Street
Boston, MA 02118
Tel: (800) 551-2121

Dynal Inc.
5 Delaware Dr.
Lake Success, NY 11042
Tel: (800) 638-9416
Fax: (516) 326-3298

Dynatech Laboratories, Inc.
900 Slaters Lane
Alexandria, Virginia 22314
Tel: (800) 336-4543
(703) 548-3889

Eastman Fine Chemicals,
Eastman Kodak Company
1001 Lee Road, Building 701
Rochester, NY 14650
Tel: (800) 225-5352
(716) 588-2572
Fax: (800) 879-4979 or (716) 722-6054

Endogen, Inc.
68 Fargo Street
Boston, MA 02210-2122
Tel: (800) 487-4885
(617) 439-3250
Fax: (617) 439-0355

Fisher Scientific
711 Forbes Avenue
Pittsburg, PA 15219
Tel: (800) 766-7000
Fax: (800) 926-1166

FMC Bioproducts
191 Thomaston Street
Rockland, ME 04841
Tel: (800) 341-1574
Fax: (207) 594-3491

Falcon Products, Becton Dickinson Labware
1950 Williams Drive
Oxnard, CA 93030
Tel: (800) 235-5953
(805) 485-8711

Genzyme Diagnostics
One Kendall Square
Cambridge, MA 02139-1562
Tel: (617) 252-7500
Fax: (617) 252-7759

Gibco BRL
Corporate Headquarters
Life Technologies, Inc.
P.O. BOX 9418
Gaithersburg, MD 20898
Tel: (800) 828-6686
Fax: (800) 331-2286

Harlan/Sprague Dawley
P.O. Box 29176
Indianapolis, IN 46229
Tel: (800) 526-3213
Fax: (317) 894-1840

HyClone Laboratories
1725 South State Highway 89-91
Logan, UT 84321
Tel: (800) 492-5663
Fax: (801) 753-4589

ICN Biomedicals, Inc.
3300 Hyland Avenue
Costa Mesa, CA 92626
Tel: (800) 854-0530
Fax: (800) 334-6999

Immunotech Corporation
P.O. Box 860
90 Windom St.
Boston, MA 02134
Tel: (617) 787-1010
Fax: (617) 787-0315

Intergen Company
Two Manhattanville Road
Purchase, NY 10577
Tel: (800) 431-4505
(914) 694-1700
Fax: (914) 694-1429

Irvine Scientific
2511 Daimler Street
Santa Ana, CA 92705
Tel: (800) 437-5706
Fax: (714) 261-6522

Jackson Immunoresearch Laboratories, Inc.
827 West Baltimore Pike
P.O. Box 9
West Grove, PA 19390-0014
Tel: (800) 367-5296
Fax: (215) 869-0171

The Jackson Laboratory
60 Main Street
Bar Harbor, ME 04609
Tel: (800) 422-6423
Fax: (207) 288-3398

Millipore Products Division
　　Bedford, MA 01730
　　Tel: (800) 225-1380
　　(617) 275-9200
　　Fax: (617) 275-8200

Miltenyi Biotec Inc.
　　1250 Oakmead Park
　　Suite 210
　　Sunnyvale, CA 94088
　　Tel: (800) 367-6227
　　Fax: (916) 888-8925

Molecular Probes, Inc.
　　P.O. Box 22010
　　Eugene, OR 97402-0414
　　Tel: (503) 465-8300
　　Fax: (503) 344-6504

Nordic Immunological Labs
　　Drawer 2517
　　Capistrano Beach, CA 92624
　　Tel: (800) 554-6655
　　(714) 498-4467
　　Fax: (714) 361-0138

Norton Performance Plastic
　　P.O. Box 3660
　　Akron, OH 44309

Nunc
　　200 North Aurora Road
　　Naperville, IL 60563
　　Tel: (800) 288-6862
　　Fax: (708) 416-2556

Oncogene Science Inc.
　　106 Charles Lindbergh Blvd.
　　Uniondale, NY 11553-3649
　　Tel: (800) 662-2616
　　Fax: (516) 222-0114

Pacific Bio-Marine Laboratory, Inc.
P.O. Box 1348
Venice, CA 90294
Tel: (310) 822-5757
Fax: (310) 215-1938

PeproTech, Inc.
Princeton Business Park
5 Crescent Avenue, G2
P.O. Box 275
Rocky Hill, NJ 08553
Tel: (609) 497-0253
Fax: (609) 497-0321

Perseptive Diagnostics, Inc.
735 Concord Avenue
Cambridge, MA 02138
Tel: (800) 343-1346
Fax: (617) 497-6927

Pharmacia LKB Biotechnology
800 Centennial Avenue
Piscataway, NJ 08854
Tel: (800) 526-3593
Fax: (201) 457-0557

Pharmingen
11555 Sorrento Valley Rd. #E
San Diego, CA 92121
Tel: (800) 848-6227
Fax: (619) 792-5238

Pierce Chemical
P.O. Box 117
Rockford, IL 61103
Tel: (800) 874-3723
Fax: (815) 968-7316

Polysciences, Inc.
400 Valley Road
Warrington, PA 18976-2590
Tel: (800) 523-2575
(215) 343-6484
Fax: (215) 343-0214

Promega Corporation
2800 Woods Hollow Road
Madison, WI 53711-5399
Tel: (800) 356-9526
(608) 274-4330
Fax: (608) 277-2516

R & D Systems
614 McKinley Place NE
Minneapolis, MN 55413
Tel: (800) 343-7475
Fax: (612) 379-6580

Research Products International Corp.
410 N. Business Center Drive
Mount Prospect, IL 60056-2190
Tel: (708) 635-7330
Fax: (708) 635-1177

Ribi Immunochem Research, Inc.
553 Old Corvallis Road
Hamilton, MT 59840
Tel: (800) 548-7424
(406) 363-6214
Fax: (406) 363-6129

Sandoz Pharmaceutical Co.
59 Route 10
East Hanover, NJ 07936
Tel: (800) 526-0175
Fax: (201) 503-6356

Sarstedt
>P.O. Box 468
>Newton, NC 28658
>Tel: (800) 257-5101
>Fax: (704) 465-4003

Sepratech Corp.
>305 North Mac Arthur Blvd.,
>Suite 100
>Oklahoma City, OK 73127
>Tel: (800) 222-0924

Sigma Chemical Company
>P.O. Box 14508
>St. Louis, MO 63178
>Tel: (800) 325-3010
>(314) 771-5750
>Fax: (800) 325-5052
>(314) 771-5757

Skatron Instruments, Inc.
>108 Terminal Drive
>P.O. Box 530
>Sterling, VA 20167
>Tel: (703) 478-5190
>Fax: (703) 478-5197

Southern Biotechnology Associates
>P.O. Box 26221
>Birmingham, AL 35226
>Tel: (800) 722-2255
>Fax: (205) 945-8765

Spectrum Chemical Mfg. Corp.
>755 Jersey Avenue
>New Brunswick, NJ 08901-3605
>Tel: (800) 772-8786
>(908) 214-1300
>Fax: (800) 525-2299
>(908) 220-6553

T Cell Diagnostics, Inc.
38 Sidney St.
Cambridge, MA 02139
Tel: (800) 624-4021
Fax: (617) 621-1420

Thomas Scientific
99 High Hill Road at I-295
P.O. Box 99
Swedensboro, NJ 08085-0099
Tel: (800) 345-2100
(609) 467-2000
Fax: (609) 467-3087

Upstate Biotechnology (UBI)
89 Saranac Avenue
Lake Placid, NY 12946
Tel: (800) 233-3991
Fax: (518) 523-1336

Winthrop Pharmaceuticals
Division of Sterling Drug Inc.
90 Park Avenue
New York, NY 10016
Tel: (212) 907-2000
Fax: (212) 551-7856

INDEX

243

Spleen, 2
Staining of cells, 36
Stimulation/starving cycles, 56

T

T cell enrichment, 15
Th clones, 53
Thymocyte co-stimulation, 80
TNFα, 79, 126, 127
TNFβ, 79
Total production of immunoglobulins, 184

TRITC, 35
Trypan blue, 6
Türck solution, 5
Two color Elispot, 193

U

Urease, 187

W

WEHI-279 cell cytotoxicity assay, 12